Culture, Mind, and Society

The Book Series of the Society for Psychological Anthropology

The Society for Psychological Anthropology—a section of the American Anthropology Association—and Palgrave Macmillan are dedicated to publishing innovative research in culture and psychology that illuminates the workings of the human mind within the social, cultural, and political contexts that shape thought, emotion, and experience. As anthropologists seek to bridge gaps between ideation and emotion or agency and structure and as psychologists, psychiatrists, and medical anthropologists search for ways to engage with cultural meaning and difference, this interdisciplinary terrain is more active than ever.

Titles in the Series

Adrie Kusserow, *American Individualisms: Child Rearing and Social Class in Three Neighborhoods*
Naomi Quinn, editor, *Finding Culture in Talk: A Collection of Methods*
Anna Mansson McGinty, *Becoming Muslim: Western Women's Conversion to Islam*
Roy D'Andrade, *A Study of Personal and Cultural Values: American, Japanese, and Vietnamese*
Steven M. Parish, *Subjectivity and Suffering in American Culture: Possible Selves*
Elizabeth A. Throop, *Psychotherapy, American Culture, and Social Policy: Immoral Individualism*
Victoria Katherine Burbank, *An Ethnography of Stress: The Social Determinants of Health in Aboriginal Australia*
Karl G. Heider, *The Cultural Context of Emotion: Folk Psychology in West Sumatra*
Jeannette Marie Mageo, *Dreaming Culture: Meanings, Models, and Power in U.S. American Dreams*
Casey High, Ann Kelly, and Jonathan Mair, *The Anthropology of Ignorance: An Ethnographic Approach*
Kevin K. Birth, *Objects of Time: How Things Shape Temporality*

Previous Publications

Bacchanalian Sentiments: Musical Experiences and Political Counterpoints in Trinidad (2008).

Any Time is Trinidad Time: Temporal Consciousness and Social Meanings in Trinidad (1999)

Objects of Time

How Things Shape Temporality

Kevin K. Birth

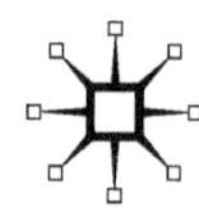

OBJECTS OF TIME

First published in 2012 by
PALGRAVE MACMILLAN®
in the United States—a division of St. Martin's Press LLC,
175 Fifth Avenue, New York, NY 10010.

Where this book is distributed in the UK, Europe and the rest of the world, this is by Palgrave Macmillan, a division of Macmillan Publishers Limited, registered in England, company number 785998, of Houndmills, Basingstoke, Hampshire RG21 6XS.

Palgrave Macmillan is the global academic imprint of the above companies and has companies and representatives throughout the world.

Palgrave® and Macmillan® are registered trademarks in the United States, the United Kingdom, Europe and other countries.

ISBN: 978–1–137–01788–8 (paperback)
ISBN: 978–1–137–01787–1 (hardcover)

Library of Congress Cataloging-in-Publication Data

Birth, Kevin K., 1963–
Objects of time : how things shape temporality / Kevin K. Birth.
p. cm. —(Culture, mind, and society)
ISBN 978–1–137–01788–8 (alk. paper) —
ISBN 978–1–137–01787–1 (alk. paper)
1. Time—Sociological aspects. 2. Time—Philosophy. 3. Time perception. 4. Material culture. I. Title.

HM656.B57 2012
304.2'37—dc23 2012006154

A catalogue record of the book is available from the British Library.

Design by Newgen Imaging Systems (P) Ltd., Chennai, India.

First edition: September 2012

Contents

Preface

First, an admission: I am merely an anthropologist who has gotten enmeshed in the study of time, and I have not been content to limit that interest to my ethnographic work. That work has been a stepping-stone into all sorts of scholarly domains where I probably have no business going, and in which I am not properly trained. My only excuse is that everyone who studies time has to choose between staying within narrow disciplinary limits or venturing outside of those boundaries. I am in the latter camp, and expect to take some lumps for it. I am admittedly in that camp because it is fun to defy boundaries. I do not claim to be a medievalist, a classicist, or a chronobiologist. I am sure that my use of sources from these fields is not as thorough as it should be, but I sincerely hope that it is sufficient to support my arguments. Even within anthropology, I cannot claim an absolute knowledge of all the excellent work done on time—much of which is embedded in book-length ethnographic writing. At a time when the field seems to reward regional specialization, I have striven to go beyond that limitation to incorporate insights gained from many parts of the world. That said, my perspective is grounded in my ethnographic experience in Trinidad, and as should become clear, that experience provides a foundation for the general points I make in this book.

This is a book about how mundane objects are used to think about time—not about the temporality of anthropology or the ontology of time. These issues are intertwined, but should occasionally be kept separate. Within anthropology, there has been a struggle to develop temporal frameworks for ethnographic representation (Birth 2008a; Fabian 2002 [1983]). There have also been arguments about the pancultural ontology of time (Bloch 1977; Gell 1992). There is much work still to be done to bring together how cognition about time is shaped, how that shaping affects anthropological representation, and the relationship between ontological ideas about time in the social sciences, chronobiology, philosophy, and physics. Those seeking answers to these big questions will not find them here, only one of my steps on my journey of finding such answers myself.

In a book that spans classical studies, history, medieval studies, anthropology, and chronobiology, the variety of citation styles is overwhelming, so I am adopting a few practices that will hopefully guide readers regardless of background. First, the in-text citations provide the date of publication for the edition I use and are followed by the original date of publication in brackets. For texts that predate the printing press, I include the date attributed to the production of the text by scholars, which, in many cases, is approximate. This prevents strange citations like "Bede 1999" and allows readers to keep the sources in context. For classical texts, I not only place the approximate date of writing in brackets, but also include within the brackets one of the standard conventions for citing classical texts by book, chapter, and line rather than by page number. This should hopefully allow those who consult translations or original-language editions other than those I used to find the quote. The bibliography, I hope, contains enough information in enough detail for anyone, regardless of specialty, to be able to find the sources I use.

With regard to my convention of dating, I continue to use the BC/AD rubric as opposed to the more recently employed CE and BCE. While the latter is more politically correct, it is also an attempt at suppressing the assumptions on which the dating system is built. The reputed date of birth of Jesus, according to Dionysius Exiguus (1844–1864 [525]), is still the anchor for the CE and BCE convention. If there is one message I strongly convey in this book, it is the value of being conscious of the logics around which our objects of time are designed. Besides, I think dating systems such as AD/BC or the Islamic system are quite special because we can point to the moment at which they were adopted, and consequently, to a moment in the past when consciousness of history was not limited to one's own group. For instance, the idea of "year of incarnation of the Lord" was first used in chapter four of Bede's *Ecclesiastical History of the English Nation* (1930 [ca. 731])—in chapter three he used a different convention. While there are those who find such systems disturbing because of their religious references, I find the ability to attach their creation to a moment in time and to a particular person or set of people important in acknowledging their historical and cultural lineages.

This book seeks to uphold a standard that I set in my previous book (Birth 2008b), namely, that rather than look upon thinkers outside of the European tradition as grist for analytic mills, to see them as offering insights that challenge the European intellectual tradition. In this case, it is Caribbean thinkers who provide the theoretical foundation for challenging the supremacy of the temporalities of globalization. This feature is not as overtly displayed at the beginning of this book as it was in my previous work, but as one proceeds through the argument, and particularly in the closing chapters, it becomes clear that my critique was instigated by Caribbean thinkers. The origins of this book are applying these postcolonial thoughts to time in the coincidental contexts of developing and teaching an interdisciplinary class on time and attending a workshop sponsored by the AHRC Culture and Mind Project at the University of Sheffield in May 2009.

Along the way, this project has benefited from the assistance of many people. John Collins, Sara Stinson, Warren DeBoer, Jim Moore, Murphy Halliburton, Timothy Pugh, Axel Aubrun, and Ted Sammons have read sections of this book. I have also benefited from conversations in person or by email with Elizabeth Hartley, Silke Ackerman, Wesley Stevens, Peter Addyman, Denis Feeney, Daryn Lehoux, Alex Bolyanatz, Jim Moore, Sara Stinson, Kate Pechenkina, Robert Canfield, and Mandana Limbert. Several people also provided me with references that I would have had little hope of discovering on my own. I thank Warren DeBoer for the references to Puleston (1979) and Kelsey (1983); Jim Moore for the reference to Brennan (1983); Ralph Hanna for helping me with Richard Rolle sources and referring me to Arntz (1981); Philip Blank for the reference to Wall (2008); and Richard Block for referring me to Rotter's experiments (Rotter 1969).

The efforts of Rebecca Lester, Brigitte Schull, and Robyn Curtis have been critical in getting this book to press in time for the end of time on Earth (whether according to the Mayan calendar or by international agreement).

Chapter five is a revised version of an article that originally appeared as "Time and the Biological Consequences of Globalization" in *Current Anthropology* 48:215–236.

The photograph on the cover and title page is of a French Revolution decimal watch, used courtesy of Stephen Bogoff, www.bogoff.com.

I owe a special debt to my family who put up with how my professional interest leaks into my daily life and even our vacations. I especially thank my wife, Margaret, who has patiently improved my thinking and my prose over the years.

Chapter 1

The Material Invention of Time

Time is invention or it is nothing at all.

—*Henri Bergson (2005 [1911], 371)*

Times have changed. I remember when I was young, whenever we changed back and forth between daylight saving time and standard time, some people would forget to change their clocks and would arrive at church at the wrong time. Now, on the Sunday morning after the time change, I sleepily gaze at my clock and then at the cable box and realize that I need to adjust the handful of clocks in my house that do not receive time signals. I no longer need to remember which weekend the time changes. An object does it for me.

Not only do objects tell time for us, and not only do some of these objects self-adjust, but we have become dependent upon objects for thinking about time. In fact, we are so dependent that it gets in the way of our imagining human societies without time-telling objects. Recently I was speaking to an audience and asked how medieval Europeans knew what time it was at night. A hand quickly shot up, and somebody blurted out, "They used their sundial." There was a moment of silence, and then some quiet giggles at the thought of somebody trying to use a sundial at night.

I asked, "What prompted you to say a sundial?"

The poor person who had offered the initial answer said, "I meant moondial, or something like that. They must have had something to tell time."

They did have techniques to tell time, but those techniques were not all mediated by objects. Some involved the observation of the movement of the stars. Somehow, our culturally shaped consciousness of time has shifted away from observing cycles in the world and toward cycles embodied in objects that are manufactured. Moreover, we delude ourselves into thinking that these objects' cycles transparently represent the rotation of the Earth and its revolution around the Sun. The truth is much more complicated, as this book explores. Yet, the purpose of this book is less to expose our cultural delusions about clocks and calendars, and more to explore the intersection of culture, mind, and objects of time and how this intersection organizes behavior.

Humans have created objects based on ideas of time deliberately divorced from terrestrial experience. Examples of this include civil calendars known to deviate from the Earth's actual rotation around the Sun, or clocks that replace the variability of the Earth's rotation with the uniformity of mechanical time. As Toda noted, time systems in organisms involve a feedback loop between the organism and the environment (1975, 318–319), but through the use of objects, humans are unusual among animals because humans create time systems with a feedback loop between the human organism and the culturally shaped world. In such a behavioral environment, as Hallowell (1955) would call it, the alarm clock has more significance than the sunrise. Moreover, objects used to tell time mediate between thought and the world in such a way that the artifact is the primary influence in this feedback loop.

The artifactual determination of time does not represent a coherent, consistent cultural system, however, but represents instead the sedimentation of generations of solutions to different temporal problems. The current ideas of millennia, centuries, decades, years, months, weeks, days, hours, minutes, seconds, and so on, are examples of such sedimented ideas—they are ideas useful in measuring durations of what happens in our world, but their relationship to our world is hardly straightforward. There are no millennial cycles in nature; the Gregorian calendar that dominates annual time reckoning has known inaccuracies; and the atomic clocks used to measure durations of a day or less are accurate to one second every 60 million years. In effect, with little reflection, one can be aware of how our time standards are a hodgepodge of different logics—some derived from our current attraction to increments of ten (decades, centuries, millennia), some derived from extreme desires for accuracy (atomic clocks), some derived from church politics mixing with astronomy (the

Gregorian calendar), and some that are anachronistic survivals of long-past societies (the choice of dividing days into 24 segments from the ancient Egyptians, and hours and minutes into 60 segments from ancient Babylonians).

Sedimentation of Knowledge across History and Culture

The previously mentioned examples of the variety of cultural logics designed into our clocks and calendars pose a challenge to how culture is conceived. Tylor's classic definition of culture emphasizes that culture is what is learned and shared as members of a society (1958 [1871], 1). Recently, the culture concept has been criticized for its potential to essentialize populations (see Ortner 2006, 12–13). In many cases, the studies of cultural concepts of time can be subjected to this criticism, because they define a society's idea of time, for example, Nuer time (Evans-Pritchard 1940) or Kabyle time (Bourdieu 1963, 1977 [1972], 1979 [1977]). In political economy, similar representations can be found, such as the construction of industrial capitalist versus precapitalist time. Such representations that attach a concept of time to a society or mode of production do not fit well with the hodgepodge of time concepts designed into clocks and calendars.

The problem of mixing ideas of time seems to apply to all cultural systems of time. One of the earliest-documented problems in the construction of a calendar is the difference between the lunar year and the solar year. A lunar month is slightly over 29 days long, so 12 lunar months total approximately 354 days in contrast to the length of a solar year, which is approximately 365 days. In Roman culture, the durations measured by water clocks, known as clepsydrae, did not coincide with hours as indicated by a sundial. Pliny the Younger's letter to Arrianus demonstrates this: "I spoke for almost five hours; for four further water-clocks were added to the twelve of the largest size already allotted to me" (Pliny the Younger 2006, 40 [ca. 100, 2.11.14]). The differences and tensions between solar cycles and work rhythms measuring duration suggest that the representation of time can be a loose collection of lunar cycles, seasonal cycles, the circadian rhythms of humans and animals, and mechanical means of ideas and perceptions and not a coherent cultural system. That said, the mediation of this loose collection of ideas by a small number of objects used to think about time can give the collection an appearance of coherence. This illusion cannot be accepted at face value, but instead, time must be approached through the study of the distribution of cultural knowledge with the mediation of such knowledge by objects playing a crucial role.

In psychological anthropology, culture has become theorized in terms of cognitive processes and models that are learned and intersubjectively shared (Shore 1996; Wallace 1961), and that shape thought. Such "presupposed, taken-for-granted models of the world that are widely shared" are cultural models (Quinn and Holland 1997, 4). Cultural models are not uniformly shared, but distributed across individuals and settings (Schwartz 1978; Shore 1996; Strauss and Quinn 1997; Swartz 1991). The focus on distributive cultural knowledge allows this perspective to serve as a foundation for exploring the inheritance and combination of logics and ideas from different contexts, settings, and historical periods.

Cultural models also vary in motivational force so it should not be assumed that their presence is sufficient to determine thought and action (D'Andrade and Strauss 1992). In the case of time, the sharing of objects of time and the logics they embody does not mean that everyone behaves in the same fashion vis-à-vis time as represented by the objects.

When treating cultural models like those of time, the issue of the distribution of culture should also take into account intergenerational knowledge and particularly what Vygotsky identifies as "the tremendously broad use of the experience of previous generations which is not transmitted from father to son" (1997 [1925], 68). As Vygotsky recognized, it is possible to study this problem through the examination of what he called psychological tools. These are "artificial formations ... directed toward the mastery of [mental] processes—one's own or someone else's—just as technical devices are directed toward the mastery of processes of nature" (1997 [np], 85). To apply Vygotsky's insights to clocks and calendars, then, involves ceasing to view them simply as technical devices that reveal time and coming to view them as tools that calculate and represent time. Then we can attend to the distribution of knowledge across objects, and to the distribution of objects across people, activities, and settings. Yet, when doing so, we must remember Quinn and Strauss's warning: "It is important to bear in mind that the mere presence today of these artifacts from the past does not guarantee their continued force" (1997, 117). But force on whom? Objects that embody cognition mediate the sharing of cognition. Continued force for the users of the representations the objects produce is different from continued force for the designers of the objects that produce particular sorts of representations.

Artifactually Mediated Cognition

Jean Lave asks, "Why does the mind with its durable cognitive tools remain the only imaginable source of continuity across situations for most cognitive researchers—while we isolate the culturally and socially constituted activities and settings of everyday life and their economic and political structures and cyclical routings from the study of thinking, and so ignore them?" (1988, 76). Representations of time are a matter of cognition and the settings of everyday life that shape cognition. Objects of time are in these settings and they are used for thinking about time. These objects are not only distributed across situations, but across cultures. Not only do the clock and calendar represent a multicultural heritage, but since they also have been coupled with processes of colonialism and globalization, they have been diffused far and wide. For instance, people in Tokyo do not use a different clock from those in Buenos Aires, just a different clock time. The transcultural continuity in temporal concepts that calendars and clocks create is one of the distinctive features of modernity. This produces a permutation of the historical issue raised about the inheritance of cultural knowledge from many traditions, namely, the distribution of cultural knowledge across many cultural traditions by means of objects. What is the nature of the transcultural continuity of ideas that objects of time create? What are the limits to this continuity? How does this transcultural continuity work to shape local cultural models to conform with seemingly global standards despite locally specific environmental cycles of daylight and biological rhythms?

To answer these questions requires an examination of the interaction of the cognition embedded in objects and the cognition of the users of these objects. What Vygotsky calls psychological tools include language and signs, and not just objects. A subset of psychological tools are things. These objects are what Hutchins calls "cognitive artifacts." Hutchins defines cognitive artifacts as "physical objects made by humans for the purpose of aiding, enhancing, or improving cognition" (1999, 126). These objects play a critical role in representing cultural models (Leont'ev 1997, 22).

Such objects can be manufactured or can simply be features selected out of the environment. For instance, in his *Georgics*, Virgil weaves together a dizzying variety of astronomical, meteorological, and environmental rhythms. To give an example, in writing about the period after the spring equinox, Virgil offers the following advice that combines plant cycles and cycles in the zodiac:

> Set beans in springtime, the time alfalfa happens in collapsing furrows,
> and millet clamours for its annual attention,
> When Taurus, gilt-horned and incandescent, gets the new year
> Up and running, and the Dog succumbs to his advance. (Virgil 2006, 13 [29 BC, 1.215–18])

Humans do not just think *about* their world, but think *with* their world. Leont'ev pointed out that "[a]ctivity necessarily brings the human into practical contact with objects that deflect, change, and enrich this activity" (1979 [1972], 52). Kirsch and Maglio explore the importance of what they call epistemic actions—"physical actions that make mental computation easier, faster, or more reliable" (1994, 513). For musicians, a common action of this sort is tapping one's foot to keep track of a beat when attempting to play a complicated rhythm from musical notation for the first time, and in fact, the use of such a counting activity has been shown to increase accuracy in time estimation (Fetterman and Killeen 1990). For Micronesian seafarers, employing the night sky as a compass and imagining their canoes as stationary while the earth and sky move past them are important methods by which they use the environment so that it allows them to meet the challenge of making landfall on tiny islands in the middle of the ocean (Hutchins 1995, 2008; Gladwin 1970).

As with the distribution of objects of time, attention to the environment involves a combination of panculturally observed phenomena, like the cycles of the Moon, and local phenomena, like the migratory cycles of birds. It also involves cultural selection from among the many observable cycles that unfold in any particular environment. In the ethnographic record, there are many examples of reckoning time by pointing to the position of the Sun. This practice, which is found in many cultures, is a model of time that stands in relationship to local cultural models, such as the Nuer relating the position of the Sun to the cycle of activities involved in caring for cattle (Evans-Pritchard 1940, 101). Such relationships between local environmental cycles and local knowledge about those cycles can become further complicated by the use of objects to mediate consciousness of cycles. For instance, in the annual ritual of shifting to daylight saving time, dairy farmers face the problem of cows not changing the time when they give milk even as market timing changes with the clock.

Whereas physical actions can assist cognitive processes about time, such actions are too labor- and cognitive-intensive for many activities. Minimally, clocks and calendars can be objects that substitute for timekeeping activities. But often cognitive artifacts do not merely substitute for thought, but actually serve to enrich it. This insight is developed in the concept of the extended mind—a theory about how aspects of the environment become coupled with the brain so as to play an active role in cognitive processes (Clark 2008; Clark and Chalmers 1998; Noë 2009). A feature of some cognitive artifacts is that they contain logics and processes that the users of the artifacts do not understand or even know. The point of using the artifacts is not to know how they work, but to assume that they work and thereby relieve their user of a particular cognitive task. Along these lines, Clark articulates what he calls the "007 principle," which is that cognitive extension allows people to structure their environments so that they know only as much as they "need to know to get the job done" (1989, 64). As Hutchins notes, "Cultural practices organize interactions with the world first by furnishing the world with the cultural artefacts that comprise most of the structure with which we interact," and that our interaction with these artifacts and the world produce cognitive outcomes (2008, 2018). Clocks and calendars are examples of this—one does not need to know the mathematics and astronomy that went into our current standards of time measurement in order to know the time; instead, all one needs to be able to do is properly interpret the output of these artifacts.

The point of most treatments of extended cognition has been to explore how the use of cognitive artifacts or the environment augments cognitive function and allows for addressing cognitive challenges that could not be carried out in one's head. The logic of augmentation is implicit in Clark's (2008) idea of the "supersized mind." Extended cognition is not unlimited cognition, however. Ratner states that "the social organization of an activity, and the cultural instruments that are utilized to carry it out, stimulate and organize psychological phenomena" (2001, 70). This combination of social organization and cultural instruments does not lead to unfettered cognition, but instead, as Leont'ev (1979 [1972]) suggested, the ability of cognitive tools to direct thought productively might also deflect and change thought. Etienne Wenger observes that such tools can focus attention and congeal abstractions (1998, 58–59), but that such tools have two negative consequences: first, the tool can come to substitute for thought, giving a false sense of understanding, such as when a formula leads "to the illusion that one fully understands the process it describes" (1998, 61); second, the tool can also give its logics a "concreteness" that they do not really have (1998, 61).

If cognitive extension supersizes the mind, then it does not do so generally, but in specific ways that are socially shaped and that reflect cultural knowledge. In effect, the social construction of knowledge, including the ways in which power can shape this construction, can be reflected in the objects by which people think. Moreover, the ability of objects to mediate power allows them to serve as a means of creating the social order through suppressing individual agency (Greenhouse 1996, 61).

The study of objects of time is the study of cognition and culture, but not of the sort limited to the mind or to a simpleminded notion of cultural boundaries. For most clock users, the logics used to determine the time are outside of their knowledge but within the objects. These logics have an artifactual existence that mediates between consciousness and the world—part of what Cole describes as the "special characteristics of human mental life" as "the characteristics of an organism that can inhabit, transform, and recreate an artifact-mediated world" (1995, 32). When one wants to know what time it is, one does not calculate it, but simply refers to a clock or watch. When one wants to know the date, one consults a calendar rather than observes the Sun, Moon, and stars. This placement of temporal logics in artifacts clearly forms a feature of humans that is quite different from anything shared with any other animal—not only do humans make tools, and not only do humans have knowledge far beyond what animals exhibit, but humans place this knowledge in tools. The cultural diversity of concepts of time is closely related to the fusion of diverse ideas and artifacts used to think. Whereas my examples so far are the clock and the calendar, the use of objects to mediate time is not new. Objects related to time are among some of the most famous in the archaeological record, for example, Stonehenge, the Aztec calendar, and the Antikythera Mechanism.

As artifacts used to think, time-reckoning tools are not merely creations of the mind, as some describe artifacts (Margolis and Laurence 2007), but creations for the shaping of thought. The incorporation of reading a clock into the curriculum of early grades is not to get children to reflect about time, but to learn to read how clocks represent time—the representation is not questioned, nor are the logics that produce it explored. The clock mediates an immense amount of complicated mathematics, astronomy, and mechanics. Its virtue is that it can package this complexity in how a representation of time is produced in a simple-enough fashion that a child can understand the representation.

Latour attributes our images of space-time to "the peculiar nature of the *objects* used in the scientific disciplines to build their measuring instruments" (1997, 185; emphasis in original). The history of innovations that lead to the time-telling representations that we expect children to master in everyday settings reads like a Who's Who of famous European scientists—Galileo, Huygens, Newton, and Poincaré to name a few. The link between the ability to precisely measure duration and the application of this ability in scientific experimentation cannot be overstated even if the resulting products are taught in early grades. In effect, time is a cultural construction, but in its current form, including its form in the grade-school classroom, it has been strongly shaped by the pursuit of scientific knowledge, *and* the key components of that construction are embedded in material objects. Yet, one wonders if as the objects shape cognition in particular ways, they diminish other cognitive skills.

Extending Cognition to Lose Understanding

In *The Art of Memory*, Yates provides an example of how cognitive extension (although she does not call it that) changes and deflects thought. In her discussion of memory techniques, including the shift from the use of mnemonic tools to literacy, one insight is that with the emergence and increasing distribution of the printed book, many of the techniques for using the environment for memory atrophied or were transformed (1966, 127)—the use of the environment as a tool for remembering diminished in the face of the information that could be stored in a book. Since the arts of memory as Yates describes them coexisted with literacy, the changes in these mnemonic arts are not the result of the emergence of literacy, but are, instead, the result of the ease with which texts could be reproduced and accessed. Ong suggests that "it was print, not writing, that effectively reified the word, and with it, noetic activity" (1982, 119). He argues that manuscripts still fostered oral use of texts—manuscripts fostered recitation and listening, whereas the mass production of texts through printing made the reading of texts a visual experience (1982, 121). The implication of the work of Ong and Yates is that certain cognitive skills that were once highly developed and common have atrophied as a result of the growing prevalence of books.

Goody and Watt's (1963) famous article "The Consequences of Literacy" heralds literacy as a significant augmentation of cognition, particularly as it relates to the storing and accumulation of knowledge. The evidence for such a claim is abundant, but the ability to retain knowledge without literacy should not be underestimated (Rubin 1995). Consequently, for all the cognitive benefits of literacy, and there are many, it becomes an example of how the augmentation of cognitive abilities in one direction deflects cognition from other practices. The myriad techniques for remembering that Rubin has explored in his experiments and that Yates discusses in her historical survey are still possible, but are not cultivated due to the greater cognitive ease of reading and writing; however, this cognitive ease—the doing of what minimally needs to be done—can lead to diminished cognitive skills and a reliance on artifacts.

A similar argument can be made about objects of time, namely, that certain cognitive skills become less common or atrophy through the reliance on clocks and calendars, even as these artifacts allow increases in precision. These skills can be cultivated, but normally are not, because their cultivation creates a sense of multiple temporalities of which clocks and calendars represent only two. To place these alternate temporalities in counterpoint with the dominant modes of reckoning time challenges these dominant modes.

Ironically, the skills that atrophy might involve the very skills originally designed into the artifact. What Hutchins says of navigation equipment and cognitive mediation is also applicable to clocks and calendars: "The symbols themselves are dematerialized and placed inside the machine, or fed to it in a form that permits the straightforward generation of internal representations. What is important about this is that all the problems the mathematician faced when interacting with a world of material symbol tokens are avoided" (1995, 363). Bargh and Chartrand suggest that this is a feature of everyday life and is not confined to such complicated tasks as navigating a ship: "In modern technological societies one encounters many such automatic devices and systems in the course of daily life. They are all devised and intended to free us from tasks that don't really require our vigilance and intervention, so that our time and energy can be directed toward those that do" (1999, 464).Through objects relieving us of the burden of the task of determining time, they also may relieve us of the burden of knowing how to determine time. Telling time by means of clocks and calendars is a mediated cognitive activity—one that links mind, artifact, and environment. Consequently, clocks and calendars are not merely tools that shape individual cognition, but are artifacts that play a central role in intersubjectively shared cognitive processes.

According to Hutchins, "The physical symbol-system architecture is not a model of individual cognition. It is a model of the operation of a sociocultural system from which the human actor has been removed" (1995, 363). In other words, the designer who worked out the algorithms designed into a "physical symbol-system" no longer needs to be present for the device to work. In effect, the design of the watch makes the watchmaker's ongoing presence unnecessary. As Glennie and Thrift specifically write about clocks: "objects are elements of networks which stress the principle that no particular kind of actor should be prioritized" (2002, 154).

An implication of this is that cognitive artifacts provide an example of how cognitive processes are distributed, reproduced, and shaped across generations and social networks that span large distances. Wertsch points out that physical objects that incorporate information continue to exist "even when not incorporated into the flow of action" and even "after the humans who used them have disappeared" (1998, 30–31). These features of cognitive artifacts have been critical to how archaeologists conduct their research, and this recognition that logics can be designed into objects can also inform how information gets distributed by the distribution of objects.

As Greenhouse writes, the exchange of commodities "potentially *transgresses* culture, to the extent that culture is presumed to be intrinsically bounded and local" (1996, 63; emphasis in original). With regard to clocks and calendars, the knowledge involved in calculating time can be embodied and represented in artifacts that transgress cultural boundaries without the majority of people who use the artifacts having to know how time is determined or who generated the calculations. Indeed, most people do not know that the time on clocks is calibrated to the cycles of cesium atoms rather than to the rotation of the Earth, have never reflected on what the "mean" in Greenwich mean time means, do not know how time is determined for time zones, do not know why the Gregorian calendar starts its year on January 1, are not aware of the astronomical calculations that went into the Gregorian calendar reforms, and would not possess either the interest or the ability to compute the complete algorithm for calculating leap years (every four years except for years divisible by 100 but not by 400).

Understanding the modern experience and conception of time requires understanding the cognitive processes that have been embedded in calendars and clocks, and these cognitive processes have moved away from phenomena directly available to human perception and toward abstract ideas of time measured and mediated by complex machines that produce intelligible results—so that charting the rising and setting of the Sun is something that humans can easily do, but the measurement of the vibration of cesium atoms is quite different. The transcultural distribution of objects of time has been accompanied by the cultivation of cognitive dependency on these objects and the logics they hide.

Cultural Appropriations of Cognitive Artifacts

It is not sufficient to document the algorithms in a cognitive artifact. For these objects to be significant in practice, they must be culturally identified and appropriated. The object, no matter how foreign its algorithms, must be used for culturally identified purposes and in culturally understandable ways. There must be cultural models for the use of the object and for making sense of its representations even if the logics embedded within the cognitive artifact are inconceivable by the user. The object must be incorporated into communities of practice. A community of practice has a social organization and its members have shared expertise and knowledge. As new people enter into the community, they learn how to interact with the other members and gradually acquire the knowledge to function as full members of the community. (Lave and Wenger 1991; Wenger 1998). Wertsch suggests that material objects that are used as cognitive tools affect their users by developing skills for "acting with, and reacting to, the material properties of cultural tools" (1998, 31). Still, one should not confuse acting with and reacting to an object with consciousness of the object's function or the logics designed into the object. Even when a cognitive artifact is used, the link between the logic in the object and practice need not be as close nor the homologies as widespread as Bourdieu (1977 [1972]) suggests. The object does not always embody the learned dispositions of thought and behavior, what Bourdieu calls *habitus*. This is particularly true in colonial and postcolonial contexts.

If one imagines a postcolonial community of practice, one gets a sense of some of the potential counterpoints between a cognitive artifact that is associated with the colonizer, on the one hand, and local practices and cultures, on the other. While the object injects its peculiar algorithms into practice, it does not determine how the results produced from those algorithms are deployed.

Because Bourdieu does not explore features of colonial domination (see Goodman 2003), his discussion of Kabyle time concepts thinly masks how concepts of time are shaped by colonial relationships, leaving one to wonder whether his description reflects Kabyle cultural ideas, or Kabyle cultural responses to French ideas. Bourdieu argues that Kabyle lack the imagination of the future and the ideas of time needed to be successful in a capitalist economy: "Free from the concern for schedules, and ignoring the tyranny of the clock, sometimes called the 'devil's mill,' the peasant works without haste, leaving to tomorrow that which cannot be done today. The alarm clock and the watch, introduced many years ago even to the countryside, do not regulate the whole of life. They merely furnish a system of reference more precise than the traditional" (1963, 58). He suggests that their rhythm of life is structured by a mythico-ritual calendar that was linked to farming cycles (1979 [1977], 27–33).

Bourdieu mentions that Kabyle are familiar with clocks and calendars, but view them as lacking utility for their way of life. His suggestion is that the logic of time represented by these devices is at odds with Kabyle rhythms. To do this, he takes the time represented by clocks and calendars as self-evident rather than as a compilation of many different logics that were imposed on Kabyle in particular ways by French colonizers. In effect, Bourdieu confuses the appropriation of a cognitive artifact for purposes of directing thought and action with the logic in the artifact.

Bourdieu does sense a relationship between objects, thought, and practice. He points out variability in the extent to which "schemes of the *habitus*" are "objectified in codified knowledge" (1990 [1980], 200). In his discussion of calendars, he is concerned with how a calendar is an objective representation that "substitutes a linear, homogeneous, continuous time for a practical time, which is made up of immensurable islands of duration, each with its own rhythm" (1977 [1972], 105). He sums up his fear as the danger of analyses that focus on "*artefacts* as impeccable as they are unreal" rather than "the uncertainties and ambiguities" that practical logic addresses (1977 [1972], 108; emphasis in original). What he does not articulate is how his own ideas are products of the codified knowledge of clocks and calendars.

The problem Kabyle had with clock time and the Gregorian calendar seems to be less a problem of the logics in these cognitive artifacts colliding with Kabyle rhythms and more a problem of French colonials using clocks and calendars to attempt to manage Kabyle rhythms. The clue is in Bourdieu's prose—Kabyle possessed clocks but did not use them to "regulate" activity (1963, 58), but instead for a precise system of references. In Bourdieu's description of Kabyle rhythms, cycles of activity are driven by proper timing in response to the weather, not by the homogeneous time of the objects of time.

Mass-produced and mass-distributed objects that embody cognition must be related to models of distributive culture, differential learning, and variations in use. What is in minds may be variable, but the variability between modern clocks and calendars is trivial. In effect, these objects come close to achieving cultural uniformity with regard to particular cognitive tasks. This objectified uniformity must be conceptualized in relationship to the variability in cognition that psychological anthropologists have been adept at documenting. The nexus of the cultural models of time embedded in an object and the cultural models of using a time-telling object need to be explored, as well as the question of to what extent the object brings about uniformity. The various examples in this book demonstrate slippage between the uniformity of cultural knowledge in the object and the knowledge of those using the objects. Indeed, there is no reason for cultural knowledge to be limited by objects, and, in fact, there are good reasons to become aware of how objects can think for us in order to consider how we wish to limit object-determined knowledge.

Sapir, Whorf, and Marx

The issues of artifacts intersubjectively shaping and constraining cognition and mediating cognitive work evokes a mixture of the Sapir-Whorf theory of linguistic relativity and Marx's theory of commodity fetishism. Sapir and Whorf proposed a theory about how language, as a tool for thought, shapes cognition; Marx offers a theory of how objects mediate social relationships. In a sense, with cognitive artifacts, both processes are happening. The objects by which people think are also objects that mediate social relations in a way that becomes obscure to those who acquire and use the objects. When combined, these two ideas suggest the possibility that some cognitive artifacts are a means by which some people can think for others without anyone being fully cognizant of it.

As Vygotsky (1987 [1934]) demonstrated, language is a cognitive tool but Sapir and Whorf suggested that since language constitutes thought, it determines thought. As Sapir wrote, "the 'real world' is to a large extent unconsciously built up on the language habits of the group" (Sapir 1949 [1929], 162). The implication is that what is expressible in one language is unthinkable to monolingual speakers of another language. According to Sapir, language has this effect on thought because it "actually defines experience for us by reason of its formal completeness and because of our unconscious projection of its implicit expectations onto the field of experience..." (1964 [1931], 128).

Timekeeping artifacts are not languages, yet they have the qualities of formal completeness that Sapir describes, and their users unconsciously project the implicit definitions of time embedded in the artifacts onto experience. One example of such unreflective projection is created by the shifts back and forth between daylight saving time and standard time in many countries. In these cases, noon remains a clock-determined noon vis-à-vis social activity, even though the Sun's highest point has been shifted by an hour relative to the rotation of the Earth. If somebody shows up at the "wrong" time as a result of the time change, then it is viewed as their mistake, not the clock's. So with time reckoning, as with language, Lucy's (1992) reading of Sapir clearly applies: "Sapir claims we anticipate (or read) experience in terms of language categories which, by virtue of their abstraction and elaboration in the linguistic process, no longer correspond to experience in a direct way" (1992, 20)—concepts of time embedded in our time-telling tools are abstracted and elaborated in a way that they no longer correspond to apprehensible cycles and rhythms in the environment, including those to which the tools were originally tied.

These ideas lead to Whorf's strong statement of linguistic relativity: "users of markedly different grammars are pointed by the grammars toward different types of observations and different evaluations of externally similar acts of observation, and hence are not equivalent as observers but must arrive at somewhat different views of the world" (1956 [1940], 221). Translated into an approach to cognitive artifacts, this means that the grammars/algorithms embedded in different artifacts that are applied to the same phenomena would produce different views of those phenomena.

The Sapir-Whorf hypothesis of linguistic relativity, as it is called, has received mixed evaluations. In his review of empirical studies of this hypothesis, Lucy (1997) has pointed out all the challenges for research—particularly disentangling claims based on content (for instance, the Eskimo, and skiers, have more words for "snow" than do other groups) from claims based on structure (for instance, monolingual Chinese speakers have difficulty thinking counterfactually because the Chinese language has no structure to communicate counterfactual statements). While there is continued fascination with linguistic relativity, the strong thesis that language determines thought to such an extent that speakers of different languages are incapable of thinking alike because of their linguistic differences has been rejected (Kramsch 2004, 239).

One reason for the rejection of the strong version of the linguistic relativity thesis is the ability of all languages to combine existing signs to communicate novel ideas, which allows for the translation of thoughts across linguistic boundaries. After all, if languages were incommensurable, would translation be possible? This criticism of a hypothesis of linguistic relativity does not apply in a straightforward way to cognitive artifacts—objects cannot be combined to create novel ideas in the same way that linguistic signs can. For instance, European monasteries in the medieval period had a means for structuring the day and the liturgy according to the canonical hours. It has been repeated often that these hours do not translate into the clock hours we currently use, but this observation also implies that the technologies used to determine time in the monasteries are incommensurable with modern clocks. To use a modern clock to divide the daylight hours into equal segments as the length of the day changes from one day to another requires a great deal of computistical ability, whereas a simple sundial does this task readily. So not only are the ideas of time different, and not only are the devices used to represent those ideas different, but representing the one set of ideas through the other technology is prohibitively difficult, and to use one technology versus the other requires being conscious of the differences in how each divides time.

This is not the result of progress or the advancement of knowledge, however. My choice of using medieval monasteries as an example here is purposeful. Medieval texts widely used as textbooks, such as Bede's *Reckoning of Time* (1999 [ca. 725]) and Isidore of Seville's *Etymologies* (2006 [seventh century]), describe days as potentially divisible into 24 equal hours, yet monasteries simply did not use such divisions until after the widespread adoption of clocks by civil society. Instead, their timing of activities was based on seasonal variations in daylight and darkness. Because they were conscious of hours of equal length, but thought it better to use seasonally variable hours, it would probably be far more difficult for the modern clock user to comprehend medieval time than vice versa.

I suggest that cognitive artifacts have far greater potential for channeling thought than language does because of the fact that objects cannot be combined the way words can. With regard to the mediation of thought and reality that fascinated Whorf and Sapir, there are artifacts that serve to mediate between thought and the world—clocks and calendars being two important examples. In effect, whereas the strength of the linguistic-relativity hypothesis is hotly contested, a parallel hypothesis about artifactual relativity has received little attention, and it holds the potential of being much stronger than its linguistic counterpart. Daniel Miller points out that material things are important because "[t]hey work by being invisible and unremarked upon" (2010, 50).

The ideas embedded in artifacts come from someone, however. The designers need not have had their thinking shaped by what they designed. The power of cognitive mediation, then, does not come from the design process, but from the relative immutability of a design once an artifact is made and distributed. It is the consumers and users of cognitive artifacts who have their thoughts unconsciously shaped. So, in fact, there are two mediations going on. The artifact mediates between the user and "reality," and the artifact mediates between the user and the designer. This latter mediation echoes Marx's idea of commodity fetishism. According to Marx, one of the distinguishing features of capitalism is the way in which the social relationship that exists between the maker and the consumer in precapitalist societies has become obscured in capitalist societies by the commodity (1977 [1867], 164–165). Commodities are purchased, and their consumers have little or no knowledge of the labor that went into designing and producing the commodity. Consequently, commodities objectify social relations through their mediation between social agents. Marx views this mediation in social, not cognitive, terms, and the ability of objects to mediate is found in precapitalist economies—the kula ring in the Trobriand Islands being a classic ethnographic example (Malinowski 1961 [1922]; Munn 1986).

In exploring Marx and Engel's theory of commodities, Appadurai raises this possibility, namely, that the production of a commodity is "the production of use value *for others*" (1986, 9; emphasis in original), but Appadurai does not explore the potential of the production of objects embodying algorithms and problem solving as use value for others. When he writes about how commodities travel great distances, and how the knowledge about them can become fragmentary (1986, 56), he does not consider what happens to the knowledge embedded in commodities with cognitive functions. Marx's distinctive insight is how commodities come to mediate and obscure the mediation at the same time. The social relations that tie producers to consumers through the commodity's mediation become obscured by the commodity. It is this feature of obscuring the mediation that I take from Marx to think about cognitive artifacts—such objects have the potential of obscuring how the objects mediate the relationship between the thinkers and those using the thoughts. In the case of clocks and calendars, their users have little or no knowledge of the designers and makers of these cognitive artifacts, much less the ideas and choices the designers made in their designs.

So not only do cognitive artifacts have the potential of constraining thought through the mediation between thought and what is thought about, but since knowledge of the artifacts' design is not necessary for their use, artifacts also hide the ways in which thought is directed and constrained. With regard to time, despite its immateriality, its representation is often mediated by material objects—whether it is stone circles, incense clocks, or electronic timekeepers: "The more humanity reaches toward the conceptualization of the immaterial, the more important the specific form of its materialization" (D. Miller 2010, 75). The specific form of the materialization of time is important for how it shapes and constrains how the immateriality of time is conceived.

Both clocks and calendars are devices that mediate between cognition and cycles in the world, but they often do so in a way in which they begin to constitute cycles that diverge from those originally observed. In such circumstances, objects of time become examples of what Searle calls self-referentiality in social concepts—but in this case, the concepts are embedded in artifacts. Searle points out that the problem with such concepts is that they are premised on belief that leads to "circularity or infinite regress in the definition" (2007, 5). The example he uses is money—money is money because it is believed to be money. He adds to this the problem that language seems to be constitutive of this self-referentiality. In the case of time, I would suggest that it is not only language, but it is the artifacts themselves that become constitutive of the self-referentiality of time. The existence of such concepts is grounded in their having functions collectively assigned to them with associated rules. In the case of clocks, the existence of the concepts embedded in the artifact then becomes somewhat separate from people, in that the clock continues to keep time without anyone using it. Every time we consult a clock, it does not need to calculate time anew, because it counts time measurements continuously.

In such cases of self-referentiality, power and authority trump accuracy. The drift of calendars away from astronomical cycles has been documented for classical Greece (Dunn 1998, 1999; Hannah 2001, 154; Pritchett and Neugebauer 1947), the Julian calendar, and the Gregorian calendar. Other cases, such as the Mayan calendar's unit of 260 days (*tzolk'in*), or Julius Caesar's choice of January 1 for the beginning of the year, have no obvious astronomical referent.

Whatever calendar one uses to think about time, the tendency seems to be for most people to refer to the calendar rather than to directly observe the skies. Calendars create cycles and even affirm incorrect dates for astronomical phenomena. An example of the latter is how the difference between the astronomical spring equinox and the calendrical equinox brought on the change from the Julian calendar to the Gregorian calendar. The spring equinox is critical to the calculation of the timing of Easter, which the Council of Nicaea in 325 AD defined as the first Sunday after the first full moon after the spring equinox. Since the period of fasting known as Lent began before the spring equinox, the equinox needed to be predicted for liturgical purposes. To ease the prediction, the Church decreed March 21 as the equinox so that all calculations could be made from that calendar date. This date was represented in terms of the Julian calendar, but this calendar is inaccurate to about one day every 130 years. Consequently, the decreed equinox drifted away from the astronomical equinox, and astronomically savvy churchmen such as Richard Grosseteste and Roger Bacon recognized the extent of this drift (North 1983, 81–83), and this led to a push for reform that, after 300 years, led to the adoption of what is now called the Gregorian calendar. But even Pope Gregory XIII's papal bull, *Inter Gravissimus* (2002 [1582]), which instituted the calendar, determined the equinox by decree rather than by demanding astronomical accuracy. Consequently, despite the corrections of the Gregorian calendar for purposes of calculating Easter, the Church's practice can still produce a date for Easter different from the principle attributed to the Nicene Council that Easter be the first Sunday after the first full moon after the vernal equinox. In effect, through calendars, power can trump observation in the determination of cycles and consequently in the activities associated with those cycles. Thus, cognitive artifacts can mediate power relations at the same time they mediate cognition.

This indicates an equivocal component of cognitive extension—the qualities by which artifacts can direct cognition could become the qualities by means of which cognition is deliberately shaped and limited. Wertsch describes the potential of cognitive artifacts as "a sort of 'taming,' or 'domestication,' of novices' actions in the world" (2007, 186), and adds, "For example, learning how to deal with a set of data from empirical observations by employing a particular graphing technique provides insight into patterns that would otherwise remain undetected, but it also entails being less able to see other patterns that could be revealed by employing different means" (2007, 186). Clocks and calendars limit cognitive models about time to those ideas that are consistent with the assumptions on which these time-reckoning artifacts are founded.

Competing Objects

The awareness of objects mediating choices is most easily seen when there are competing objects. Clocks contrast with sundials—but that is a competition clocks won long ago.

Probably the most interesting example of competing clock systems was during the French Revolution after metric time was adopted (Zerubavel 1977). In this system, the day consisted of

10 hours of 100 minutes each. This system lasted for only a short time, but during that period some very interesting clocks and watches were produced. All of them not only display metric time but the more conventional 12-hour or 24-hour cycle, as well. An example of such a watch is on the cover of this book. If you examine the image closely, you'll realize that the time of 12:52 on the conventional dial is equivalent to the time of @0:37 on the metric dial. The clock was designed to do the conversion for its user rather than relying on the user to convert in his or her head.

Today, the most notable competition is between calendrical systems. About calendars, P.-J. Shaw is justified in stating:

> A date is the symbol of a moment rather than the moment itself, and a calendar is a device for identifying a day, month, sometimes a year, distinct from a system of reckoning, which is a tool for computing the passage of time. But because the modern (Christian) calendar acts also as the modern system of reckoning and is universally acknowledged as such, the correspondence between day and date, between a moment and its given symbol, is so close that the two tend to be treated as identical. One consequence of this is that the artificial nature of that date becomes obscured; it assumes the privilege...of a universal law. (Shaw 2003, 29)

As with clocks, the Gregorian calendar conflates the reckoning of points in time with the measure of duration. Feeney comments on the Julian calendar: "[A]s inhabitants of Caesar's grid we take it for granted that a calendar is there precisely to *measure time*, to create an ideal synthesis of natural and socially or humanly organized time and in the process to capture a 'time' that is out there, waiting to be measured, but the product of the operation of measurement" (Feeney 2007, 194; emphasis in original). The Gregorian calendar's emphasis on measurement over timing is not found in all other calendrical systems. Some calendars, such as the Hindu calendar, emphasize timing (Good 2000). Even the Gregorian calendar has anachronistic elements that emphasize the relationship of celestial cycles, such as the determination of the timing of Easter (although this is done by tables approximating celestial events rather than by astronomical observation).

In a society such as Trinidad, in which there are large Christian, Hindu, and Muslim populations, the complex interaction of different calendars with different emphases is very apparent, and this creates an elaborate social polyrhythm of holidays (see figure 1.1). Some holidays are fixed in the Gregorian calendar, like New Year's Day and Christmas. Other Christian celebrations, like Easter and J'Ouvert (Carnival), move. J'Ouvert moves because it is tied to the beginning of Lent, and consequently to the timing of Easter. Phagwa and Diwali are Hindu holidays; they occur on the same day every year in the Hindu calendar, but when transferred into a Gregorian rubric, they move. Eid and Hosay are Muslim holidays, and are fixed within the Islamic calendar, but like Hindu holidays, when transferred into a Gregorian calendar, they appear to change their dates from year to year.

There is a subtle structuring of figure 1.1 that I wish to make explicit. The year used to structure the placement of all the holidays of all the religions is the Gregorian, that is, the Western European Christian year. These holidays could be represented within the Hindu annual cycle, or within the Islamic year, but the choice of the Gregorian year was conscious—that is how these days are represented within Trinidad. So while the artificiality of

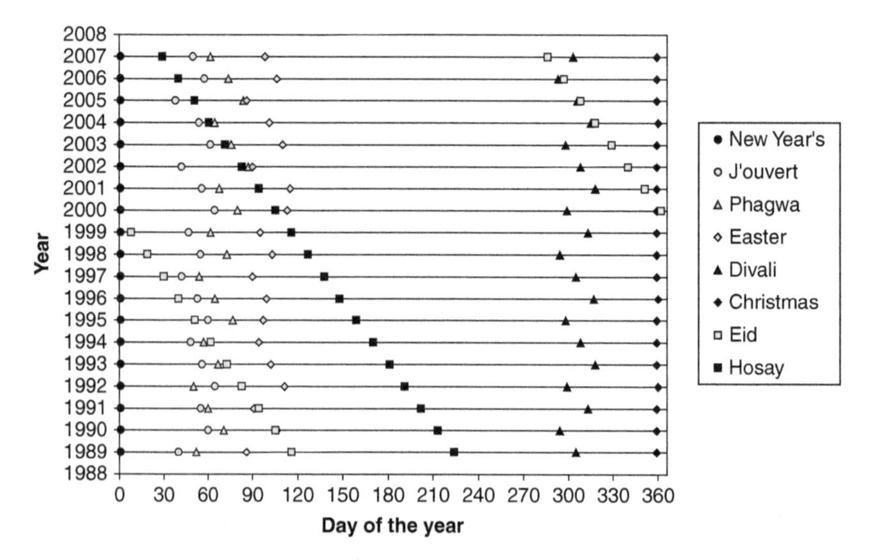

Figure 1.1 Polyrhythm of holidays in Trinidad by day of the year, 1997–2007 (from Birth, forthcoming).

the Gregorian year might be more clearly conscious than that of clock time, that consciousness has not diminished its influence or the tendency to subsume all other calendars under the Gregorian and to represent their timing in Gregorian terms. It does mean that the Gregorian calendar functions a bit differently than the clock—rather than hide competing temporalities, it absorbs them and potentially eviscerates them of their logics in favor of their critical moments and cycles being marked as Gregorian dates. In the case of both clocks and the Gregorian calendar, so much cognition is encoded in the artifact that it is difficult to consider alternative temporalities without first converting them into clock and Gregorian representations, but in this action of translation, much information is lost. This makes it difficult to think about not only alternative temporalities, but temporalities in general.

Artifactual Determination of Time

I have argued that the concepts of time embedded in the cognitive artifacts used for time reckoning shape our cognitive processes about time so much so that it becomes a challenge to think about time in any way other than what the artifacts dictate.

One way to test this claim is to look at communities of practice where clocks or clock-like devices lead their users to misjudge duration or timing. Such cases demonstrate the extent to which users trust and defer to the devices even when they are in error.

Despite the large literature on time perception, very little has been done on whether altering the rate of a clock can affect cognition and perception of time. Instead, the literature explores how accurate humans can be at estimating clock time. In itself, this suggests the power the clock now has over how psychologists think about time.

There are some psychology experiments in which the speed of a clock was altered to explore whether it affected the time judgments of research subjects. The unequivocal results were that covert alterations in the tempo of a clock did not diminish research subjects' attribution of veracity to the clock (Craik and Sarbin 1963; Rotter 1969). Rotter's (1969) experiments demonstrated the extent to which a clock could be slowed and research subjects still give credence to its indication of time. In his experiment, subjects were given passages to read and were told to record the time by using a laboratory clock for which the rate of its motor could be adjusted without the research subjects knowing. Rotter chose four clock speeds—100 percent of normal, 50 percent, 25 percent, and 12.5 percent. The subjects wrote down the time at the beginning of each passage, and without their subjects knowing, the clock speed was adjusted downward after they made their record of the time. After 32 minutes, the subjects were asked to use a seven-point scale to rate their interest in the readings, the rate of the flow of time, and whether they thought the hands on the clock were moving more slowly or quickly than normal (1969, 46–47). The results were that subjects perceived the clock as moving faster than normal when it was actually running at a normal speed; they perceived the hands moving at a normal rate at 50 percent clock speed, and time flowing as normal at 25 percent of clock speed. Moreover, Rotter concluded, "it appears that while *Ss* [research subjects] were expressing judgments and feelings about clock-speed, they accepted the clock readings as being veridical" (1969, 48). In effect, in a conflict between how subjects sensed time and the time indicated by the clock, the subjects treated the clock as giving the correct time.

Latour (1997) is suspicious of the division between subjective and objective time, and these experiments give credence to this view. Instead of a divide between internal time senses and external time cues, it seems that time concepts are a mediator between the environment and the mind, and do not reside solely in one or the other.

Rotter's study was conducted before the widespread availability of quartz clocks and clocks that received radio signals from the international time grid. The variability in watches was such that the *New York Times* reported in 1928 on how the United States Bureau of Standards had a division that tested watches, and the report also addressed why all mechanical watches tended to deviate from accurate timekeeping (January 8, 1928, 124). Variation in clock times among clocks and watches was part of daily life—yet, even with the cultural awareness of the variability in clocks, the power of the clock trumped the subjective perception of time.

Even now, in 2012, the ability of clocks that run slow to affect how their users reckon time has also become a component of a popular computer-based practical joke in which the joker loads an application on the victim's computer to alter the rate of that computer's clock. Supposedly, great amusement ensues when the victim misses appointments as a consequence of consulting the time on the computer.

It is possible that errors in time reckoning that result from reliance on cognitive artifacts have existed as long as the artifacts themselves—the earliest evidence I know of such an error is associated with the third century BC in Rome. Pliny the Elder states of a public sundial brought from Sicily to Rome: "The lines of this sundial did not agree with the hours, but they were followed for 99 years, until Q. Marcius Philippus, who was Censor with L. Paulus, placed a more precisely constructed one next to it" (Pliny 2005, 106 [ca. 77–79, 7.214]).

The social consequence of this was that in a community of practice, different individuals could think it was a different time at the same time. This, in fact, happened. Good evidence comes from diaries and journals kept by individuals engaged in the same endeavor at the same time in which they recorded the time. The example I shall focus on is Sullivan's expedition against the Iroquois in 1779 during the American Revolution. This expedition is remarkable for the period because of the large number of extant soldiers' journals that are easily available.

Sullivan's expedition began in the summer of 1779. It combined several regiments of regular soldiers with a few militia units. Its ostensive purpose was to punish and neutralize the Iroquois who, the summer before, had attacked the settlement at Wyoming along the Susquehanna River, decimating its militia force and seriously compromising the ability of settlements in the area to defend themselves. Another purpose of the expedition was to map and survey this fertile region. The campaign mostly involved the Continental army destroying Native American settlements and farms. There was only one major battle between the two forces—occurring outside of present-day Elmira, New York. The campaign not only succeeded in militarily neutralizing the Native Americans as allies of the British, but also in weakening them to the point of allowing for settlement, and the survey and mapping information gleaned greatly assisted in the expansion of Euro-American farmers.

Sullivan often issued orders in terms of clock time. Lieutenant Colonel Francis Barber of the 3rd New Jersey Regiment recorded these orders in a surviving order book (Murray 1975). The rank-and-file soldiers did not necessarily know the time associated with the marching orders but simply followed the commands of their officers or responded to the signals played on drums. Several of the journals record events in terms of clock time. When the orders found in the different journals are compared, there is considerable variation in the times recorded.

During the eighteenth century, the watch had become an increasingly common accoutrement of the military officer, and by the end of the century, officers in most European armies were required to carry watches (Sauter 2007, 700). In *Discipline and Punish*, Foucault uses the military as an example of embodied rhythms. One of the points of military drills and discipline was to create a shared embodiment of movement such that troops moved at the same time in the same way as a coordinated mass (Foucault 1977, 162–169). The order book from Sullivan's campaign provides evidence that implies that clocks were used to coordinate maneuvers—particularly the mustering and marching at the beginning of each day. The orders were then communicated to the rank and file through sonic signals. Indeed, in the one extant order book, in the record for August 4th, one finds the following:

> The following signals are to be established, to wit:—
> Two ruffs will be a signal for the whole to march in files
> One ruff to march in single file
> Three ruffs to advance in platoons
>
> The Troop being beat upon the march is ever the signal for the column to close and
>
> Beating To Arms A signal for displaying unless special orders be given to the contrary at the time.
>
> In order that no mistake may take place in the signals, an orderly drummer or more is to be appointed in each regiment and the signal to be taken from the front and repeated through the whole line. (Murray 1975 [1930], 59)

I shall focus on September 4th, 1779, and I do so for two reasons. First, on the evening of September 3rd an unusual set of orders was given for the next day's march that necessitated reliance on clock time. Second, according to all the journals, September 4th was a rainy day, which precluded the use of the Sun as a means of reckoning time.

Possibly prompted by the engagement at Newtown just a few days prior, or by the sighting of an enemy scout on September 3, in the order book kept by Lieutenant Colonel Barber, one finds the following: "The General to beat tomorrow morning at five o'clock at half an hour after the Assemblee, and at three-quarters after the march without the usual signal" (Murray 1975 [1930], 88). This is corroborated by the journal of Lieutenant William Barton, who wrote on the evening of September 3rd, "This evening orders were given to march at half past 8 in the morning, without the usual signals" (Cook 1887, 9); and Lieutenant Erkuries Beatty who wrote on the morning of September 4th, "Recd. orders last night to march today 5 oClock without the usual Signals of Guns firing" (Cook 1887, 28).

According to Baron von Steuben's regulations for the Continental army, the beating of general is the signal to strike the tents and load the wagons; the beating of assembly is the signal to form into battalions for marching; and this is usually followed by the signal to march (von Steuben 1779, 67–68), although the orders for September 4th give a clock time to start marching and employ the use of a duration, rather than a sonic signal, to begin the march.

Not surprisingly, the journals of that day give no clear picture of when the army marched. Just as the records of the order for marching differ in the time the march was to begin, the accounts of when the army actually moved vary. For instance, the three officers in Spencer's 5th New Jersey Regiment from whom we have accounts recorded the time of march that day quite differently: Major John Burrowes indicated the time of march as nine in the morning, Dr. Jabez Campfield recorded it as eleven, and Sergeant Thomas Roberts recorded it as ten.

The difference between the orders issued the night before and the time of march could be attributed to many factors having to do with mustering on a rainy day, but that is not the only discrepancy. As a result, the variation cannot be simply attributed to differences created by the order of march. Instead, the variation suggests several things. First, consistent with Glennie and Thrift (2009), the accuracy of clock time was not of great concern even after watches and clocks became common personal items. Second, even within an army on a campaign, the need for accuracy in time reckoning was not great during this period. Third, there was no strongly felt urge to get times to agree in the written record, or to get clocks to synchronize. Finally, the time was what an individual thought it was. As Alexander Pope wrote in his *Essay on Criticism*, "'Tis with our *Judgments* as our *Watches*, none / Go just *alike*, yet each believes his own" (1970 [1711], 4; emphasis in original). If watches within a regiment differed in their time the way opinions differed, then whose time was correct? The answer seems to be that it did not matter.

"What, then, is time?"

Augustine of Hippo wrote, "What, then, is time? If no one asks me, I know; if I want to explain it to someone who asks me, I do not know" (1997 [ca. 397–98], 256). The question stumps us for quite different reasons. We surround ourselves with cognitive artifacts to tell us what time is, and the time these artifacts represent is demonstrably confused in accuracy. We have hyper-accurate atomic clocks and a calendar that poorly represents the duration of the Earth's orbit. The nature of time's existence is confused by cognitive artifacts and by the human invention of the time constructs these artifacts indicate.

If one adopts Aristotle's perspective that time is the measure of motion (1936 [ca. 350 BC]), then a system that is as inaccurate as the Gregorian calendar cannot reflect true time, but this is not the source of much, if any, anxiety. This lack of concern about the existence of the time with which we organize our lives seems to be part of the modern condition as distinct from the concerns of the predecessors of modernity. For instance, Alliney says, in his introduction to the questions of William of Alnwick on time, that the ontological status of time was a more significant topic in fourteenth-century Europe than in our own time (Alliney 2002,

XL). Alliney is only partially correct in this statement, in that Einstein's theory of relativity makes the relationship of time and motion a central concern, but Einstein's theory is relegated to physics and denied entry into social scientific and literary theory treatments of time—as Hassan and Purser note, it "barely dented the modernist assumptions we [social scientists] had internalized regarding how we viewed time" (2007, 8). Derrida, the seeming master of textual relativity, explicitly dismisses temporal relativity. In writing about one of Baudelaire's short stories, Derrida states, at no given or desired moment "can one reasonably hope to find, *outside any relativity*, noon at two o'clock" (1992, 34; emphasis added).

It is ironic that Derrida, the arch-postmodernist, chooses absolute temporal uniformity, and that the physicists adopt relativity. The ideas of time important for daily life and the construction of knowledge are dependent on objects, and the presently used objects—modern calendars and clocks—are relatively recent in their form and design. Their synchronization across different contexts is still more recent. Whereas there are cognitive benefits from the precision derived from the measurement of small durations, and there are also cognitive benefits in unburdening the mind from having to calculate the time, it is also the case that these devices have channeled cognition in specific ways. When not recognized, this channeling, I argue, constrains our ability to understand time across cultures and to ascertain temporal characteristics of our world not subject to the clock and calendar. By deferring cognitive processes to these objects, we run the risk of diminishing our ability to think about time, and we also run the risk, as Greenhouse states for clocks, of using the objects as "a materialization of some universal time sense" (1996, 7).

Timescapes

Crucial to breaking free of the constraints of time tools is the recognition of timescapes in relationship to communities of practice. Adam developed this concept as a means of thinking about the way in which multiple temporalities and rhythms interact in complex ways. She writes, "I suggest that we think about temporal relations with reference to a cluster of temporal features, each implicated in all the others but not necessarily of equal importance in each instance" (2004, 143). These temporal features and relations can include biological processes, environmental cycles, and cultural constructs (Adam 1998). In effect, Adam gives a name to the target of LeFebvre's concept of rhythmanalysis—a methodological emphasis on "temporalities and their relations within wholes" (2004 [1992], 24). Clocks and calendars are artifacts that emphasize particular temporalities and rhythms, but they are merely powerful contributors, as opposed to determinants, of the timescapes in which they are found. This book explores how clocks and calendars function as cognitive artifacts, but then moves to look at these artifacts in the context of other cultural processes that organize cognitive ideas and embodied experiences of time.

Another crucial step to break free from these constraints is the use of ethnographic and historical data on timescapes to challenge the dominance of the cognitive artifacts currently used to think about time. This is important for the recognition of how transcultural continuity of objects of time produces variable practices. As Greenhouse notes, the dominance of these objects of time has tended to create a view of the reckoning of time as either being done through the use of artifacts like those employed in the European technological tradition, or as being closely associated with observations of nature, and she adds, "when ethnographers look for time in societies without the West's timekeeping technologies, they look for it in cultural knowledge about natural cycles. Interestingly, some of them do not find it there" (1996, 40).

It is not enough to demonstrate alternative temporalities—anthropology and history have done that with great skill. Instead, the coexistence of multiple temporalities with the time of clocks and calendars must be shown. Gupta (1994) and Chakrabarty (1997) push this possibility, but since they do so with a European/non-European dichotomy, they still leave the exclusiveness of European time reckoning intact. I shall pursue a different strategy—a counterpoint of disrupting clock and calendar time from within European traditions coupled with ethnographic challenges to European clock and calendar temporalities from the postcolonial and colonial context of Trinidad. Since the

Caribbean represents the incubator of European imposition of homogeneous, uniform time on non-European laborers (Mintz 1985, 51), it also represents the place of long-term non-European temporal dissent.

In the study of time, it is important to fend off the disciplinary divides that conveniently keep physics, chronobiology, philosophy, psychology, and social science from impinging on one another. As Latour writes of his work, "this research does not deal with nature or knowledge, with things-in-themselves, but with the way all these things are tied to our collectives and to subjects" (1993, 4). Time is a domain that spans disciplinary boundaries—a subject without limits on which limits have been imposed. Whereas it might seem dizzying, and even unsound, to move between ethnographic data, physics, chronobiology, and historical material on the American War for Independence as well as historical and archaeological material on classical and medieval Europe, I think each move is justified. Each body of material raises questions that often require turning to other sources of information.

Since time brings together so many different domains of knowledge and experience, reducing time to what our cognitive artifacts tell us is an unfortunate and unnecessary constraint. This is not a book antagonistic to these devices, however, but a book that seeks to reveal their limits, so that we use these artifacts in moderation with careful consideration.

Chapter 2

A Necromantic Device, or How Clocks Think

Time should be so defined that the equations of mechanics may be as simple as possible. In other words, there is not one way of measuring time more true than another; that which is generally adopted is only more convenient.

—*Poincaré, from "Time and Its Measure" (1913 [1905], 227–228; emphasis in original)*

Clocks and calendars are necromantic devices—tools by which the dead think for the living, and the dead's thoughts deflect the living's attention from the cycles in the present. This is a consequence of the mediation of cognition by artifacts, and it is a feature of how artifacts can distribute cognitive models across time, culture, and space.

What are the origins of the cognitive models programmed into our objects of time? The contemporary Gregorian calendar was the product of a Vatican-appointed committee (Ziggelaar 1983, 205–209) that reformed the calendar imposed on the Roman Empire by Julius Caesar (Feeney 2007). Clocks as we know them represent multiple conventions of international time standards preceded by the invention of mechanisms for reliable chronometers (see Landes 1983; Cipolla 1978; Sobel 1995); by the invention of the pendulum by Galileo (Dohrn van-Rossum 1996; Landes 1983, 116–118); by the concept of the day being divided into 24 hours of equal length (the earliest reference to these is third century BC; see Hannah 2009, 124–125); and by Roman, Greek, Egyptian, and Babylonian concepts of dividing the day. So, then, clocks and calendars mediate networks of distributed cognition consisting of great historical depth, transcultural efforts, and many different institutions—and since most of the cognitive contributions to these objects were made by people who are long dead, but whose ideas have been programmed into clocks and calendars, these cognitive artifacts are, in a sense, means by which the dead think for the living.

Most clock and calendar users do not view these artifacts as necromantic devices—the extent to which the dead think for the living is obscured by the artifacts' mediation between their users and their designers. Not only are the original thinkers behind the clock's and the calendar's algorithms hidden from the consciousness of their users, but the assumptions and decisions on which the algorithms were created also remain hidden. This prevents the users of these artifacts from fully recognizing the dilemmas and alternatives the long-dead designers of these time-reckoning tools addressed in their work.

This chapter discusses one particular cognitive tool, the clock, and examines the ways in which the dead thinking for the living constrains the latter's thinking about time. The next chapter will discuss calendars. The argument here will follow three paths. The first is to shake the ontological complacency with which clock time is viewed. The second involves the use of ethnographic examples to reveal how clock time can be determined without the use of clocks. The third path of argument will examine medieval European concepts of time to reveal alternative temporalities that predate the clock. I choose to focus on medieval Europeans for two reasons. First, in the literature on the emergence of time consciousness, the representation of medieval Europeans has played a role in drawing the contrast between capitalist, clock-measured time, and precapitalist societies that claimed to have no time consciousness (e.g., Anderson 2006 [1983], 22–36; Le Goff 1980; Taylor 2007, 58–59; Thompson 1967). Second, there are many primary sources from medieval Europeans that address time—Stevens estimates over 9,000 extant medieval manuscripts about cycles of time (1995, 46). Such documents are not filtered through an ethnographic perspective cognitively shaped by the clock.

The three paths of argument are meant to demonstrate the possibilities of time outside of clock-mediated methods of determining time—a case of a tool coming to shape, if not to determine, cognition. This point has methodological consequences of suggesting that much of the social scientific approach to time has been subtly mediated by the models of time inherent in clocks rather than by attending to the many features of timescapes. Even more broadly, as pointed out in the previous chapter, it suggests a hypothesis that bears similarities to the Sapir-Whorf hypothesis of linguistic relativity and connects it to Marx's idea of commodity fetishism at the same time, but instead of language determining thought, it is a class of objects that determines thought and obscures a process of mediated distributed cognition.

What Is the Time?

Clocks are tools that tell us the time, but that obscure the answer to the question of what time is. In fact, it is interesting that the conscious discussion of the existence of time in the European intellectual tradition typically reaches conclusions that are at odds with the clock. As Einstein put it:

> A clock is a thing which automatically passes in succession through a (practically) equal series of events (period). The number of periods (clock-time) elapsed serves as a measure of time. The meaning of this definition is at once clear if the event occurs in the immediate vicinity of the clock in space; for all observers then observe the same clock-time simultaneously with the event (by means of the eye) independently of their position. Until the theory of relativity was propounded it was assumed that the conception of simultaneity had an absolute objective meaning also for events separated in space.
>
> This assumption was demolished by the discovery of the law of propagation of light. (1992 [1926], 382)

Einstein's concept of the space-time continuum is jarringly divergent from NASA's practice of using clocks set to Coordinated Universal Time (set to the meridian of Greenwich, England) as the time standard for space shuttle flights regardless of their position and velocity. Bergson's critique of the spatialization of time and its division into discrete units (2001 [1913], 75–139) contradicted the time zones that were adopted during his own life; and recently, in physics, Lynds (2003) has challenged the independent existence of time. Basically, clocks cognitively naturalize assumptions about time that are challenged by those who contemplate the existence of time.

The referent of clocks is not what most clock users think it to be, namely, it is not solar time. This creates a disjunction between the semantic and the pragmatic aspects of clock time. The semantic referent of clocks is the reliable cycles of cesium atoms in atomic clocks distributed across the globe that are regulated by the Bureau International des Poids et Mesures (International Bureau of Weights and Measures, abbreviated as BIPM). The common cultural attribution of what clocks indicate—the pragmatic meaning assigned to them, or the interpretant in Peirce's terms (1955, 99)—refers to an idea that the clock is tied to the position of the Sun in the sky. From a semiotic perspective, the dominant cultural interpretation of clocks fosters a delusion.

The components of this delusion and its development can be summarized as follows. The popular understanding of clock time held by many clock users is that it measures the rotation of the Earth. The Earth is thought to spin on its axis once every 24 hours. So an hour is simply 1/24th of this rotation, a minute 1/60th of an hour, and a second 1/60th of a minute. This understanding has been unfounded since the 13th General Conference on Weights and Measures in 1967—a conference at which cesium clocks were chosen as the preferred timekeeper over any astronomical cycle. Since that conference, the model of time within our clocks is no longer tied to the calculation of the rotation of the Earth, but is merely reset on occasion with leap seconds to bring it back in accord with the Earth's slightly wobbly, irregular rotation. Even that practice may now end based on the recommendation of the Radiocommunication Sector of the International Telecommunication Union (ITU-R), which sets time-signaling policies under the United Nations, and with it, any reference to the Earth's rotation in the measure of time. Newton noted the problem with the use of the Earth's rotation as a timekeeper: "For the natural days are truly unequal, though they are commonly considered as equal, and used for a measure of time; astronomers correct this inequality that they may measure the celestial motions by a more accurate time" (1934 [1687],

7–8). In effect, time as indicated by the heavens and the rotation of the Earth is too irregular for science.

This frustration with the wobbly Earth prompted the search for a means of measuring time that approached Newton's desire for an abstract, accurate time with which to measure phenomena, and consequently, to construct knowledge. In this search, there was a move away from the use of astronomy to reckon time, and toward independent, absolute time (Ariotti 1973). Initially, the noncelestial measure of time was related to the movement of a pendulum, but this proved too irregular, as well (Ariotti 1973, 50); this led to time being set to atomic clocks. In effect, pivotal scientists have pursued an abstract time divorced from natural events, and while this idea of time is now foundational in the post-Enlightenment's construction of knowledge, most users of clocks are unaware of clock time's disconnection from the rotation of the Earth.

The measurement of atomic cycles is not the only cognitive model embedded within the clock. In another move that suppresses temporal relativity in favor of simultaneity, the Earth is divided into geometrically unjustifiable yet politically convenient time zones. In theory, these are at intervals of 15 degrees of longitude, which represent a difference of one hour between each time zone and its neighbors to either side, but some countries, like Venezuela, choose to be a half hour off their neighbors, and other countries, like China, prefer to have a single time for the nation even though it should span multiple time zones. Still other places create time zones, like Kiribati, which is 14 hours ahead of Greenwich, England. A map of the world's time zones looks like a jigsaw puzzle rather than a division of the globe into 24 equal segments. Most owners of clocks and watches do not know the longitude to which the time of their time zone is calibrated, or the politics about time zone definitions. Instead, they simply note when they cross time zones, and adjust their clocks and watches accordingly.

Daylight saving time is still another political decision that is part of current models used to reckon time. The change between standard time and daylight saving time is increasingly programmed into clocks and watches so that their owners do not need to know when to reset their timepieces. The idea of daylight saving time was originally proposed by Benjamin Franklin, but was only adopted in the early twentieth century (see Prerau 2005). For many years, it was the responsibility of clock and watch owners to heed public announcements of when changes occurred between daylight saving time and standard time and to adjust their timepieces. Increasingly, this has become unnecessary as electronic devices that tell time are adjusted automatically. Recently, I heard a tale of a group of people waking up the morning after the change back to standard time, consulting their cell phones to learn the time, and then over breakfast debating what was wrong with the clock on the kitchen wall, which was suddenly one hour fast. As the person who told me the story admitted, it took her some time to realize that it was the day of the change (she had not remembered initially), and that the kitchen clock was not broken, but needed to be adjusted manually. As self-adjusting devices become common, one wonders if the back-and-forth change between daylight saving time and standard time will begin to escape people's consciousness except for their wondering each spring why they are awaking when it is dark, when the day before it was light, and why each fall it is suddenly unusually dark when they leave their workplace at the end of the workday. With the extent to which modern cognitive processes have been trained to ignore environmental time cues, it is possible that even these clear temporal indications of light and dark might escape notice.

There is also more to the algorithms in the clock than how time is reckoned. Clocks actually embody and combine two distinct cognitive processes—processes that I shall argue were separate in the European Middle Ages—namely, clocks determine both what the time is at a particular moment, and the measurement of the length of the durations involved. In fact, clocks subordinate the determination of moments and timing to the measure of duration. Midnight is no longer halfway between sunset and sunrise—instead sunrise is represented as a duration that has passed since midnight. So, for instance, in Boston on August 22, 2006, the Sun set at 7:35 PM, but rose the following morning at 6:00 AM: midnight was 4 hours and 25 minutes after sunset, and 6 hours before the next sunrise.

The merger of ideas and the subordination of timing to the measurement of duration is foundational to the modern construction of knowledge. Time is a case of what Latour and Woolgar describe as phenomena that are "thoroughly constituted by" material instrumentation (1986, 64).

Many fields of knowledge rely on methods that involve the measurement and conceptualization of time, but I shall briefly discuss only economics. In this field, the ideas embedded in methodologically useful artifacts, such as clocks and calendars, become assumptions that underpin the theory. The classical economic theory of the value of labor, particularly Marx's elaboration on Adam Smith, is a set of ideas on which many authors have based their discussions of the time consciousness of industrial capitalism—"Marx's determination of the magnitude of value implies a sociohistorical theory of the emergence of absolute mathematical time as a social reality and as a conception" (Postone 2003, 218). His clearest statement of the use of time to represent labor is in *A Contribution to the Critique of Political Economy*, where he writes: "Just as motion is measured by time, so is labour by labour-time. Variations in the duration of labour are the only possible difference that can occur if the quality of labour is assumed to be given. Labour-time is measured in terms of the natural units of time, i.e., hours, days, weeks, etc." (1970 [1859], 30, emphasis added).

Hours are culturally constructed, not natural (Dohrn van-Rossum 1996); and the same is true of the idea of the week (Zerubavel 1985). In effect, Marx treats two cultural constructions as natural units of time. Labor-time becomes measured and theorized in terms of these units.

This idea of labor-time is not the time of a particular laborer or a particular act of labor, but an abstraction. According to Marx, "Labour, which is thus measured by time, does not seem, indeed, to be the labour of different subjects, but on the contrary the different working individuals seem to be mere organs of the labour" (1970 [1859], 30), and "[t]he same labour, therefore, performed for the same length of time, always yields the same amount of value, independently of any changes in productivity" (1977 [1867], 137). Postone extrapolates from Marx's discussion of labor-time that "Marx's determination of the magnitude of value...implies that time is an independent variable, the homogeneous, absolute mathematical time that has come to organize much of social life in our society, has been constituted socially" (2003, 216). This abstracted labor-time then leads to a source of tension between what Postone calls concrete time and abstract time (2003, 292), but even the idea of concrete time embedded in Postone's analysis involves the productivity of a particular worker in a particular period of time, and not a challenge to the unit of measurement.

This idea of labor-time is not limited to Marx. The use of time to represent standardized labor was important in the methods of scientific management that emerged from Frederick Winslow Taylor's work on "scientific management." Taylor's concept of science involved measuring the amount of work performed in uniform durations to create standards and incentives to increase efficiency (Kanigel 1997; Taylor 1967 [1947]). In Taylorism, time, motion, and work-pace studies were used to create a standard for evaluating workers—how much a properly trained worker could produce in a set amount of time. His idea was that proper management would result in worker productivity, and, consequently, worker earnings would increase (Taylor 1967 [1947]).

This logic was taken even further by Becker's inclusion of the allocation of time as a variable in economic analysis (1965). In an implicit extension of Taylor's logic, Becker states that "[m]ost of the large secular increase in earnings, which stimulated the development of the labour-leisure analysis, resulted from an increase in the productivity of working time due to the growth in human and physical capital, technological progress and other factors" (1965, 505). But Becker not only discusses the increased efficiency of production, but also of consumption (1965, 506). In effect, the classical economic means of representing labor in terms of uniform durations becomes a factor in representing consumption, as well, and the relationship between consumption and production becomes theorized in terms of the cost of time.

The use of time as a standard of measure and a means of creating comparisons between different activities relies on a view of time as uniform, and consequently, it results in a suppression of the importance of timing. Concepts such as quarterly or annual earnings hide the importance of the holiday shopping season or agricultural cycles. Hourly wages obscure the reality that workers are busier some hours than others.

While there have been some attempts to bring timing into economic theory (Hamermesh 1999; Winston 1982), the tendency is to erase the timing of concrete events altogether in favor of representations of abstract duration. For instance, as Shackle wrote of economic theorists' tendency to imagine static systems in equilibrium: "[W]e do not seek to put time in fetters or rob it of its essence but, instead, we abolish it altogether" (1967, 94). And as Danby has more recently written in a criticism of neoclassical economics, "*material processes take place in time*... It is only by a truly heroic act of assuming this reality away that neoclassical general equilibrium can model all of this temporally organized activity as a timeless process of simultaneous market clearing" (2004, 56; emphasis in original).

Thus, the merger of different modes of time reckoning and the dominance of the measure of duration have become hidden assumptions not merely of time, but of the construction of other forms of knowledge. The cultural and historical specificity of clocks has not undermined the confidence their users have in applying them universally. The devices themselves have come to shape our thinking about time, even as it is easy to criticize how clocks represent time. To understand how these devices work, I seek to loosen the grip of these necromantic devices over our thoughts—to explore alternate temporalities and ways in which clocks mediate that are different from how we are accustomed to using timekeeping objects.

Clock Time without Clocks

Clocks are not a necessary condition for time reckoning—Bilfinger (1892) gives extensive examples of references to time from medieval Europe, and more recently Glennie and Thrift (2009) demonstrate for the late Middle Ages in England that clocks did not enable the reckoning of time but were adopted by a society that was already engaged in practices of reckoning time. Moreover, one should heed Sorokin and Merton's criticism that "[m]ost social scientists have proceeded on the tacit assumption that no system of time other than those of astronomy or the imperfectly related calendar is possible or, if possible, useful" (1937, 615), and be open to nonastronomical and noncalendrical ways of conceptualizing time. Before I explore medieval time reckoning, I am going to take what may seem to be an unusual approach and begin with questions that emerged during my field research in Trinidad, where I encountered many examples of determining clock time without clocks. I shall offer three brief examples.

These examples turn the typical mediation of clocks on its head—instead of clocks indicating the timing of a cycle in the environment, these are cases of time cues independent of mediation by the clock. The cues discussed are not the normal ones associated with time, like solar or stellar cycles. This suggests that clock time need not come from a single knowledge base, but can be at the intersection of many kinds of experience and many kinds of knowledge.

The first example is an exchange I had with a friend while walking down the road in rural Trinidad. We were headed to a junction in the road where a group of men often gathered around dusk—a fruitful place for an ethnographer to be each evening. As my friend and I were walking, we heard this loud cacophony of bird calls as a large flock of parrots passed above us. My friend turned to me and said, "It must be six o'clock—the parrots are flying overhead." This observation struck me—this young man was determining clock time without a clock. The daily routine of birds was sufficiently accurate for him to know the time.

The second case involves an interview that I had arranged for six o'clock one evening with a middle-aged Creole man. As I approached his house, he greeted me and said, "You late—the boys done play small goal long time!" Every evening a group of young adult men would gather on a field in front of this man's house to play a version of soccer that involved a small field and small goals. Typically, they finished playing by around six in the evening. For some reason, this particular evening they had finished early. The man I was scheduled to interview admitted to me that he did not know how to read a watch, but that he was quite capable of telling time based on what was going on around him. The small-goal game typically finished at six, so my arrival long after the game had finished must mean that I was late for our six o'clock appointment. Like the observation about the parrots, this is a case in which clock time was reckoned by something other than the clock.

The third instance comes from an interview with an elderly Indian man. I asked him about his use of clocks and watches, and he said:

> Well, most times, I for one never own a wristwatch. I own a clock, but not a wristwatch. Now, I only use clock at mornings mostly. The few times I really make time important to me are in the morning, like I want to leave for work at a certain time, and through that, I will always gaze upon the clock...But if I am in the bush, at a certain time, I does have that feel it could be so and so time, and it does be around that very time. Sometimes, I would know, I will say, now is about three, right? Sometimes, you trying now to reach a certain point you might estimate it's three, and it's four, and you have to go to a funeral so you have to come home to bathe, but you want to reach and you know it's three o'clock, so you put back that piece of work, cuz you don't want to left it anyhow to come back. You see, most of the times we look at the weather, how it looks. We look at the Sun, imagining that does be a source of time. It might not be exact, but it usually is not too far.
>
> KB: What if it is a cloudy day?
>
> A: You still have this thing, once you move within certain time and you working. You see, if I left home without checking what time I leave, I would be in a little problem, but once I go there, every time I almost am within time. I estimate and sometimes I just be—you have that in you. Sometimes on work they will ask you, and you will say now is about, it could if it's not nine, it can't be too far past. It is only about three or four minutes past. You have that in you.

The vignettes and interview material offer tantalizing insights. One is that clocks are not the only indicators of time to rural Trinidadians. In effect, clocks are not necessary tools for reckoning time. This point has been well documented in ethnographic literature, with Evans-Pritchard's discussion of the Nuer cattle-clock as one of the most famous cases (1940, 101–102). Atkins offers another example from Zululand: "it was the music of beasts, birds and insects that engaged the immediate attention for the singing or calling was kept up the whole twenty-four hours constituting a day, by various animals, in turn, as their *time* for performing came round" (1988, 238; emphasis in original).

What makes the three Trinidadian cases different from the Nuer cattle-clock described by Evans-Pritchard or the beast-based time reckoning of Zululand is that Trinidadians use environmental phenomena and their own time sense to determine a clock time. This was not a case of a system of time reckoning divorced from clock time, but of environmental rhythms sufficiently repetitive to be useful in determining clock time without a clock. A second implication of this material is that capitalism and the emergence of clocks has not erased other forms of time reckoning. The contemporary Caribbean's social systems were forged by capitalism (Mintz 1985, 1993), and Trinidad, specifically, has been a site for ethnographic discussions of the relationship between globalization, capitalism, production, labor, and consumption (Yelvington 1995; Miller 1994, 1997). The drawback of my ethnographic material is that it still refers to clock time, which makes it only a tantalizing indicator of cognitive processes of time reckoning without the clock. Ethnographic research in Trinidad is conducted in a setting in which references to clock time are ubiquitous, and, as the example of the parrots demonstrates, references to clock time occur even without a clock.

In one sense, it is reasonable to point out that one would not know that the parrots flew overhead at six o'clock if one did not have a clock, but on the other hand, what of the possibility that the parrots could serve as a time signal independent of clocks? How would such a system of time reckoning work? For one thing, parrots are not sufficient to reckon time throughout the day, but only indicate a particular moment during the day. For another, their behavior, while useful in determining a particular moment, would not be useful in measuring duration. Nonhorological time reckoning could not rely on parrots or soccer players alone. In this respect, it would differ radically from the use of the clock—clocks provide a single time reference, but nonhorological thought, if parrots and soccer players are guides, must weave together a variety of time indicators, some of which are only useful for determining the time of day, but not for measuring duration, and many are only useful at a specific time of day.

This would work like Linnaeus's flower clock. In his *Philosophia Botanica*, he writes that, because of the daily blooming cycles of certain flowers, it is possible to create a flower clock—a garden by which one could reckon time according to which blossoms were open and which were closed (2003 [1751], 295–297). Using such a floral clock would require more than looking at a single flower, but would necessitate ascertaining the distinctive combination of open and closed blooms for each hour of the day. In effect, at different times, one would pay attention to different flowers. But even the flower clock refers to the clock. To examine how time was reckoned before the clock, one can turn to the abundant textual evidence remaining from the European Middle Ages.

The Importance of Time in Medieval Europe

In the tale of progress about the emergence of clock time, medieval Europeans are often represented as being unable to determine time. Marc Bloch wrote that "[t]hese men...lived in a world in which the passage of time escaped their grasp all the more because they were so ill-equipped to measure it" (1961, 73), and that "the imperfection of hourly reckoning was but one of the symptoms, among many others, of a vast indifference to time" (1961, 74). Even when it is granted that at least monastic communities kept time, the representation of that time is "irregular" (Thompson 1967, 71–79) or "careless of exactitude" (Le Goff 1980, 44).

The claim that medieval Europeans were indifferent to time has no evidence to support it. Putter comments that "the impression of 'medieval indifference to time' is only the reflex of our indifference to medieval time" (2001, 136). Le Goff points out that once we accept that the reckoning of time in the Middle Ages was different than now, "far from being indifferent to time, men in the Middle Ages were singularly sensitive to it" (1988, 175).

This look at medieval European sensitivity to time is to note the objects and elements used to think about time during that period in that society. In so doing, one quickly recognizes that this involves recognition of the distribution of cultural models across settings and communities of practice. Unlike modernity, in which the same objects are applied in many different settings, medieval European cultures relied on setting-specific and problem-specific objects and techniques. Medieval science includes examples of the ability to judge duration, such as Bede's study of the timing of tides (Stevens 1995, 37), or the use of applied astronomy in monastic timekeeping (McCluskey 1998, 99ff). The representation that medieval Europeans had no sense of time is undermined by the attribution of the origin of the concept of the hour to medieval European monasteries (Borst 1993, 26–32; Dohrn van-Rossum 1996, 29–43; Landes 1983, 61–62) and to these monasteries' role in public time signaling (Bilfinger 1892; Glennie and Thrift 2005, 171); the suggestion that medieval Europeans lacked time awareness is also challenged by the popularity of the celebration of lay versions of the monastic offices. As Glasser observes, "The most effective argument against the alleged absence of a sense of time in antiquity is that arbitrary subdivision of the day, the *hora,* which the Middle Ages inherited from antiquity" (1972, 55). These "little hours" were daily cycles of recitations of prayers, litanies, psalms, and canticles that were tied to the signaling of the canonical hours by local churches and monasteries. They were signaled with a variety of bells that served as indicators of time to the public (Bilfinger 1892; Glennie and Thrift 2009, 83–85).

The popularity of lay liturgical celebrations associated with these hours is indicated by the fact that the single most prevalent type of manuscript to survive from the Middle Ages is the breviary for lay people. These books guide the celebration of the daily liturgical cycle, and after the printing press was invented, the production of these breviaries far outnumbered the Bible (Weick 1988, 7). Duffy describes the *Book of the Hours* as "the most popular book of the late Middle Ages" (2006, 4). This suggests that the time signals applied to monasteries and churches mediated knowledge of time for lay people, as well.

Even though the printing press and the clock emerged in the same period, the daily liturgical cycle remained tied to canonical hours, not to clock time. This remained true long after the widespread use of clocks. Corbin suggests that this difference persisted in some parts of France until the late nineteenth century (1998, 110–115), and Bruegel demonstrates that it persisted in the Hudson Valley of New York until the beginning of the nineteenth century (1995, 548–549), which is evidence that these variable hours were socially significant enough to be reproduced through processes of colonial expansion and migration. Such far-flung persistence supports Glennie and Thrift's (2009) argument that the power that clock time had over temporal organization during the eighteenth and nineteenth centuries has been exaggerated.

The representation of medieval time as "irregular" is also misleading. This is based on the conflation of uniformity and regularity. The contrast between these concepts can be understood in musical terms. Both a steady beat and syncopated polyrhythms can be cyclical, predictable, and regular, and both can offer a temporal structure to a song. Only the steady beat has uniform intervals, however. To suggest that prehorological time was irregular is basically tantamount to suggesting that many musics in the world built around polyrhythmic structures are also irregular. Consequently, a more accurate portrayal would be to place the regularity of medieval European timekeeping at the level of layered repetitions of cycles and to give it the same attention that ethnomusicology has given to various rhythmic genres. Medieval timekeeping was therefore more complicated, not more irregular, than modern timekeeping by clock time that relies on uniformity.

The so-called irregularity is due to the length of the hours being tied to the amount of daylight—something that varies considerably with the seasons and latitude. Bede described these variations, and his works on time reckoning became widely distributed throughout Europe and served as textbooks. With regard to seasons and latitude, Bede wrote:

> Traveling through the southern zone in wintertime, it [the Sun] rises earlier and sets later for those who inhabit the southern regions of Earth than [it does] for us who, placed towards the north, with the globe of the Earth blocking the way, receive its rising later and its setting earlier. But on the other hand, in summer

> the Sun rises very much earlier for us who live under the same latitude, and seems to remain much longer on the point of setting [with us] than with those who dwell on Earth's southern flank. (1999 [ca. 725], 92)

These differences were part of a regular annual activity cycle institutionalized by Benedict of Nursia's monastic rule (1998 [ca. 530], chapters 8, 10, 41, 47, and 48), which became the model for monastic routines throughout Europe.

Such differences made the concept of the hour more of a boundary between periods of time than a duration. In his *Etymologies*, Isidore of Seville applies his method of emphasizing homonyms to indicate the definition of "hora"—"*Hora* is a Greek word, though it sounds Latin. For *hora* is the limit of time, just as *ora* [the Greek word ὥρα] is the edge of the sea, rivers, or clothing" (2005 [sixth to seventh century], book V, chapter 29). This idea—indeed this specific phrase or a variation of it—was often repeated, appearing almost word for word in texts about time by Bede (1844–64b [ca. 703], col. 278–279; 1999 [ca. 725], 15), a dialogue attributed to Alcuin (1844–64a [eighth century or later], 1113–1114), and others. Alcuin, a Northumbrian cleric who became a leading figure and educator in Charlemagne's court, wrote: "Siquidem et in nocte stationes, et vigiliae militares in quatour partes divisae ternis horarum spatiis secernuntur" [And even in the nighttime stations, and the soldiers' watches are divided into four parts by means of three hours separating the spaces of time] (Alcuin 1844–64b [eighth century], 1274). If one imagines three points in time, then one can come up with four watches—the watch before the first point, the watch between the first and second points, the watch between the second and third points, and the watch after the third point—four parts divided by three hours. This confusion is ameliorated by Glasser's observation that in ancient French, *heure* was used primarily to indicate a "point in time" and had "no definite, more or less precise durational significance"—it was not used to refer to duration until it gained use in Middle French (1972, 56). Buddenborg points out that this way of conceptualizing the hour creates confusion—it is often difficult to discern whether an hour refers to the point at which it begins, or sometime between hours, or a completed period from one hour to the next, and he suggests that the basic principle in St. Benedict's Rule was to assume that an hour referred to the completed hour unless explicitly stated otherwise (1936 [ca. 530], 90). Consequently, while it was possible for hours of uniform durations to be imagined, in European practice before the clock, it was not, and a consequence was a separation of the determination of the time of day from the measurement of duration.

If the use of technologies that measure duration are a guide, then the cognitive task of measuring duration was separated from the determination of a time of day. An examination of how water clocks (also called clepsydrae) and hourglasses were used indicates that they were used solely to measure duration, as in the following riddle from Symphosius:

> Lex bona dicendi, lex sum quoque dura tacendi
> Ius avidae linguae, finis sine fine loquendi
> Ipsa fluens, dum verba fluunt, ut lingua quiescat
> [Hard rule of I, good rule of speech
> To words that know no end, an end I teach
> I flow as well as they, that rest the tongue may teach.] (1912 [fourth to fifth centuries?], 47)

Even after the widespread adoption of the clock, hourglasses were used to measure stable durations—John Donne makes reference to them as determining the length of sermons (Wall 2008).

Contrary to Le Goff's argument that merchants and employers "felt the need of *measuring* more exactly the time of work and of commercial operations" (1988, 182–182; emphasis in original), measurement was not necessary for medieval merchants to know when transactions took place. Merchants kept track of times of day in reference to variable canonical hours—Epstein (1988) provides a study of this practice by thirteenth-century notaries in Genoa. To reckon time in the Middle Ages was fairly easy for merchants since the ringing of bells in monasteries and churches to indicate these hours was part of the urban soundscape. But since the length of canonical hours varied with the seasonal changes in the amount of daylight, these hours could not be standards for measuring durations (Bilfinger 1892, 142ff).

There is a danger in viewing the clock as necessary for certain cognitive tasks simply because we use it for those tasks. The importance of clock time in twenty-first-century economic practices cannot be used as grounds for assuming that it was necessary for economic practices in the medieval period. This misconception is key to the view of medieval timekeeping as irregular or even nonexistent. As Rothwell points out, the use of clock time to represent the canonical hours of the European Middle Ages distorts how time was reckoned, and "is at the root of most misunderstandings about the measurement of time in the Middle Ages" (1959, 241). This should give some cause for concern about the assumption that there was a lack of time consciousness in the Middle Ages because no clocks were used, and this also raises the very real possibility that the time consciousness of the Middle Ages is simply not translatable into clock time—or put another way, time consciousness in the Middle Ages is not embedded in the logic of design of our modern clocks. This lack of uniform clock time is not a lack of an awareness of time. On the contrary, before the clock both informal and disciplined activities relied on ideas of timing and time derived from a variety of environmental and liturgical cues (Glennie and Thrift 2002, 159)—liturgical cues were publicly significant because they were often signaled by bells and adapted to secular purposes, as will be discussed later. How time was reckoned in medieval Europe in comparison to industrial Europe reveals a shift in cognitive process away from the perception of multiple indicators of time that needed to be reconciled, and toward modern horological thought that unquestioningly relies on clocks to measure duration to indicate time.

Chaucer's Timescape

The *Canterbury Tales* are full of references to time reckoning in a period that predates the widespread adoption of clocks, and a period that viewed the reliability of clocks with skepticism. Chaucer's time reckoning also is comparable to Trinidadians' because of the use of multiple cues for time that could even indicate a clock time without reference to the clock. What Glennie and Thrift state for premodern England fits with what I have described for Trinidad: "Any event with a known starting or finishing time (church services, civic processions, market activity, working days, leisure events, and so forth) provided temporal information merely by happening" (2002, 165).

Chaucer was an expert on time reckoning (North 1988; Mooney 1993), having penned a treatise on how to use an astrolabe to determine the time (Chaucer 1988a [ca. 1391]). This knowledge of time reckoning was manifest in his more famous work, *The Canterbury Tales*. For instance, in the prologue to the Man of Lawes' tale:

> Oure Hooste saugh wel that the brighte sonne
> The ark of his artificial day hath ronne
> The ferthe part, and half and houre and moore
> And though he were nat depe ystert in lore
> He wiste it was the eightetethe day
> Of Aprill, that is messager to May;
> And saugh wel that the shadwe of every tree
> Was as in lengthe the same quantitee
> That was the body erect that caused it.
> And therfore by the shadwe he took his wit
> That Phebus, which that shoon so clere and brighte,
> Degrees was fyve and fourty clombe on highte
> And for that day, as in that latitude
> It was ten of the clokke, he gan conclude. (1988b [ca. 1390], 87)

In this passage, time is determined not by reference to a single device, but by perceiving the position of the Sun, looking at the length of shadows, and relating those perceptions to knowledge about one's latitude and the date. This is a much more complicated cognitive task than simply looking at a watch, and at its heart is not only observation, but seeking corroboration between different sources of knowledge, and reconciling those sources. Indeed, Mooney argues that Chaucer and his contemporaries "were in the habit of citing the time by several methods consecutively, as if to compare them" (1993, 92).

In a discussion of the challenges faced by medieval navigators of around Chaucer's period, Frake documents the multiple aspects of the timescape that skilled sailors used to tell what the state of the tide was in a location, and the direction in which it was flowing. He mentions that modern sailors rely on the cognitive artifact of tide tables for this information, but that medieval sailors did not have such tools (1985, 259). For medieval sailors, time of day was determined by estimating solar time using the compass rose. Solar noon was when the Sun was at its highest point, and at that point it was due south for these North Atlantic sailors. At night, time was estimated by the position of the stars of the Little Dipper as they circled Polaris. The compass rose is divided into 32 points, which are used to divide the day into 45-minute periods. Moreover, the points of N, NE, E, SE, S, SW, W, and NW indicated to the hours of 12:00, 3:00, 6:00, and 9:00 AM and then 12:00, 3:00, 6:00, and 9:00 PM. In this way, time could be reckoned in relationship to the Sun. The challenge then was to relate this to the lunar cycle, which determined the tide. According to Frake, in northern Europe, at lunar noon, the Moon is also due south. The compass rose could then be used to determine the time of lunar noon in relationship to solar time. In effect, the tidal patterns of a location were represented in terms of the lunar time of high tide in relationship to the compass rose's indication of solar time. The medieval navigator, then, was conscious of the relationship of the cycles of Moon, Sun, and tide, and how each was charted through the use of the compass rose. In contrast, the cognitive tools of tidal tables, clocks, and GPS, or even just software that uses GPS information to indicate the local tides, are cognitive tools that obscure the extensive cognitive tasks involved in determining one's position in time and space before the advent of these technologies.

Chaucer's time reckoning was not limited to observing the skies. In the Nun's Priest's tale he tells of a rooster that was more accurate than a clock:

> Wel sikerer was his crowyng in his logge,
> Than is a clokke, or an abbey orlogge.
> By nature he crew eche ascencioun...
> Of the equynoxial in thilke toun;
> For whan degrees fiftene weren ascended,
> Thanne crew he, that it myghte nat been amended. (1988b [ca. 1390], 253–254)

The crowing of roosters was viewed as an accurate means of determining midnight during the Middle Ages (Birth 2011b). In fact, the Christmas midnight Mass was called *missa gallicantu* in Latin—the Mass of the rooster's crow. In a discussion of rising to pray before dawn incorrectly attributed to the fourteenth-century English mystic Richard Rolle (see Arntz 1981), there is a reference to the cock crowing before sunrise: "quickly rise from thy bed at the bell-ringing: and if no bell be there, let the cock be thy bell" (Rolle 1910 [fourteenth century], 116). Moreover, the intelligence and wisdom manifested by roosters in telling time was held up as a model for clergy—in the sixth century, Gregory the Great wrote in his *Morals on the Book of Job*: "The cock also received understanding, first to distinguish the hours of the night season, and then at last to utter the awaking voice" (1850 [sixth century], 370). *Morals on the Book of Job* was one of the most widely read books in the Middle Ages (Gilson 1955, 108), and Gregory the Great's use of roosters as analogies for priests was often quoted, cited, or echoed throughout the Middle Ages. In his book on birds, Hugo de Folieto wrote of the moral wisdom of the rooster, and its ability to discern the hours at night, and he explicitly named his source as Gregory's *Moralia* (1844–64 [twelfth century], col. 33). Willene Clark describes this book on birds as "a teaching text for monastic lay-brothers, using birds as the subjects of moral allegory" (1982, 63). With the growing accuracy of clocks, roosters' reputations seem to have suffered, and they have been transformed from keepers of time during the night, to simply quaint signalers of sunrise, or even outright nuisances.

Multiple Time Indicators and Work

Chaucer's writings indicate that time was not determined by reference to a single device or even to just astronomical observation, but that the determination of time involved seeking multiple cues that needed to be reconciled. Despite his emphasis on the irregularity of time reckoning in medieval Europe, Le Goff does note that "[i]n daily life, medieval men used chronological points of reference borrowed from different sociotemporal frameworks" (1988, 177). Timescapes can vary with environments, and a difference between the medieval and modern timescapes is that in the modern timescapes, all cycles are related to a limited set of widely shared and intersubjectively significant cognitive artifacts—the clock and calendars—whereas in Chaucer's timescape, there were many sources of information on time that created context-dependent models for time reckoning; for example, the cock is useful for determining the predawn hours, but not for structuring day labor.

Chaucer was not unusual. The complexity of medieval timescapes in Britain can be inferred from many sources—the writings of Bede (1999 [ca. 725], 1844–64b [ca. 703], 1844–64a [early eighth century]) and Isidore of Seville (2005 [sixth–seventh century]) were extremely influential and used by Aelfric (1942 [ca. 993]) and Byrhtferth (1995 [ca. 1010]) in their texts instructing others on how to reckon time. Basically, the understanding of time was part of formal instruction throughout the Middle Ages. Moreover, Byrhtferth's *Enchiridion* was meant to instruct parish priests (Baker and Lapidge 1995, lxxix), which meant that knowledge about time was not limited to monastic communities, but was expected to be woven into the rhythms of local communities. Discussions of liturgy in relationship to time of day and season was a common element of monastic rules (Knowles 1951, 1966, 450–456). Discussions of daily life also reveal temporal complexity (Holmes 1964; Neckam 1863 [ca. 1180]).

To demonstrate this, I shall focus on a document from the Fabric Rolls of York Minster in which labor and timekeeping are both discussed. This document outlines the workday for construction workers in 1352. The Black Death had struck York in 1349 and had devastated the population for about a year. One of the consequences of the plague was a shortage of skilled workmen. The result was not a simple manifestation of the law of supply and demand, but instead a complex set of statutes that emerged throughout Europe without any clear pattern other than governments being motivated by an anxiety of increased power among laborers (Cohn 2007). In England, the power and wage pressure exerted by the shortage of skilled labor led the King's Council to act in 1349 and Parliament to issue a statute in 1351 to control wages. In her study of these labor laws, Putnam states that "their main object was to secure an adequate supply of labourers at the rate of wages prevailing before the catastrophe" (1908, 3). Maybe this control over wages is why the negotiation of time and work breaks, as opposed to wage increases, seems to be such a feature of the post-plague Fabric Rolls. As a result, the emphasis on time not only represents the time consciousness of those seeking to have York Minster Cathedral built, but also represents the time consciousness of the workers doing the construction, and their insistence on generous meal breaks with mid-workday naps, and time for drinking at the lodge.

The representation of the workday begins with a discussion of when work begins and when the first break occurs:

> Quod ipsi cementarii carpentarii et ceteri operarii incipere debeant operari singulis diebus operalibus in estate usque ad pulsacionem campanae B.M.V.... [That the masons, carpenters, and other workers should begin working at daylight on each working day during the summer, until the striking of the bell of the Blessed Virgin Mary...] (York Minster 1859, 172; all commas removed from the Latin transcription since they do not appear in the original Fabric Rolls' text)

There are several things to keep in mind in this passage. First, in recent practice, daylight is represented as sunrise/dawn and referenced to as clock time, but from the time of the Roman Empire through the Middle Ages, the period from night through sunrise was divided into several distinct segments, and predawn through morning liturgical practice was tied to this division (Knowles 1951, xv–xvi; 1966, 450–451). To link these periods to clock time creates a false uniformity across seasonal variations and latitudinal differences in daylight, and it deflects attention from the distinctive qualities of different periods of the night that were noted by medieval communities of practice and influenced the cycles of activity for some of these communities, such as night watchmen and cloisters.

The first relevant part of the night is *gallicinum*—"quando gallus resonat" [when the cock crows] (Bede 1844–64b [ca. 703], col. 281). This is actually before dawn and was associated with midnight: in his *Etymologies*, Isidore of Seville describes the rooster's crow as "the midnight blast, the announcement of day" (2005 [sixth—seventh century], (book V, chapter 30). It is important to remember, however, that "midnight" does not mean an exact clock time, but literally, a period during the middle of the night. The first crow of a rooster marked the end of the midnight period. This was the time of the first prayer service of the daily office as practiced by the clergy, the time known as nocturnes, but the modern association of cockcrow strictly with dawn has led to some confusion.

Following *gallicinum* was *matutinum.* In his *Etymologies*, Isidore of Seville defines it as "between the withdrawal of darkness and the arrival of dawn" (2005 [sixth—seventh century], book V, chapter 30). This is not morning in the sense of a period after sunrise, but is instead the first light that precedes the Sun's appearance. In the *Regularis Concordia*, a tenth-century agreement seeking consistency in practice among English monasteries (Knowles 1966, 42–48), the celebration of "Matutinales Laudes de Omnibus Sanctis" [Lauds for All the Saints] is followed by "Laudes pro Defunctis" [Lauds for the Deceased], and ends at daybreak—"Quod si luce diei ut oportet finitum" [Which, if it ends at the light of day, it is proper] (*Regularis Concordia* 1953 [tenth century], 15). Day truly begins with *diliculum*, or dawn, the period during which the Sun starts to rise above the horizon.

Because of these gradations and because they were signaled with bells, it is possible to imagine how workers could rise and be at work "on time." Light is too dim to work during *matutinum*, so the period between the signal of matins and the sufficient daylight for work was probably more than sufficient to get from home to the workplace. Moreover, the emphasis on work beginning at daylight suggests that the workday began long after the daily liturgical cycle, and somewhat independent of this cycle, as well, since the workday was not tied to the ringing of a canonical hour, but instead to the appearance of the Sun. While it comes from a period several centuries before the document describing the 1352 workday, Aelfric's *Colloquy*, a dialogue written to teach Latin, has *diliculum* as the time when the oxen are taken to the field, that is, the beginning of the workday (Aelfric 1947 [ca. 1000], 20).

One challenge posed by York's timescape is that there are many cloudy days. This means that the coordination of labor could not be solely dependent on the Sun. There might have been agreement on how to determine sufficient daylight to begin work using visual tests similar to those in Islam and Judaism for determining the timing of the first prayer of the day. In Islam, according to prehorological Islamic tradition, the morning prayer is to be recited before the Sun rises above the horizon (*Qur'an* 50:39), and its time was determined by the ability to distinguish a black thread from a white one (*Qur'an* 2:187). Judaism has a similar tradition for the morning recitation of the Shema, which, according to the Talmud (*Mishnah Berakhot* 1948, 48–49 [ca. 200, 1.2]), is determined by the ability to distinguish a blue thread from white one in the predawn light.

Based on Glennie and Thrift's (2009) research, there should be no illusion that clocks solved these problems. The period's clocks were notoriously inaccurate (Cipolla 1978 [1967], 7; Landes 1983, 83), and had to be reset according to the Sun. The continued importance of the Sun for setting clocks and watches resulted in many early watches coming equipped with a sundial and compass (Landes 1983, 88). So even once public clocks and their signals came to be widely used, their use would have been subjected to the same corroboration and cross-checking that previous time-reckoning methods would have used.

Since the timing of breaks could not be determined reliably with reference to the Sun in a place with so many cloudy days, this is possibly why not a single break time mentioned in the York Minster document refers to a position of the Sun. The time of the first break during the summer was signaled by the bell of the Blessed Virgin Mary—this was the bell associated with the daily Mass dedicated to Mary. This breakfast break of the workers seems to be more or less coordinated with a common breakfast break for monastic communities. In those communities, after prime there was a lay Mass, and after that was a breakfast break. The end of that break was signaled by the bell for the "Lady-Mass" (Crossley 1962, 85). York Minster had no associated monastic community and it is not clear that its priests had the same timing for its liturgical day as nearby St. Mary's Abbey, although Bilfinger suggests that parish churches closely followed monastic time reckoning (1892, 67). Regardless, the daily routine of abbeys was part of the timescape that could be copied or relied upon for time cues in the general community—the abbey walls did not contain the sound of its bells. Whatever the reason, then, the activity cycle of the workers was directly tied to an emulation of the time reckoning associated with a monastic activity cycle.

Whatever the exact timing and sources of the bell of the Blessed Virgin Mary, from the perspective of distributed cognition, the use of this bell to determine the breakfast break is different from the start of the workday, which was not signaled by a bell, but was determined by the amount of daylight. It suggests a timescape that included both visual and sonic cues. Daylight was something that could be observed directly, but the means used to determine striking the bell of the Blessed Virgin Mary is something that those working on the construction of the cathedral deferred to whomever rang the bell. Since the purpose of the bell was to signal part of the daily liturgical cycle, the bell was not meant to structure the workday, but instead was an intersubjectively recognized time signal upon which the workers and their supervisors agreed. Even though the bell was not for the organization of construction labor, it still mediated cognition between the workers and the bell ringers.

The length of the breakfast break was not determined by the position of the Sun in the sky or bells. The Fabric Rolls of York Minster state: "et tunc sedeant ad jantaculum infra logium fabricate cum non jejunaverint per spacium dimidiae Lente:" [and then they should sit down to breakfast in the lodge of the works for half a space, if they are not fasting during Lent] (York Minster 1859, 172). "Half a space" refers to a measurement of time. *Spacium* was commonly used in Latin to refer to a period of time; the term also refers to a "circuit." Whatever the actual duration to which *spacium* refers, by virtue of its being tied to the time it takes to walk a particular distance, it would not vary with the seasons in the same way as the duration between the canonical hours varied.

The use of travel distance to estimate time is widespread. For instance, for the Asante, the measure of time and distance were intertwined. A measure of time was the rhythm of a person's normal pace, and a measure of distance was how far a person could walk in a day (Wilks 1992, 179). The Asante kingdom was then conceptualized in terms of day-long walking distances measured from the capitol (1992, 181), with the largest sphere of Asante influence being viewed as a circle with a diameter of 40 days (182–183). The conceptual relationship of travel time and distance was also a fundamental feature in Polynesian navigation (Hutchins 1995, 65–93). In ancient Judaism, the concept of *mil* referred to both space (about 2,000 cubits) and a duration based on walking (S. Stern 2003, 54).

In medieval England, there are two techniques that could be used to measure a uniform space of time that are clearly documented. The first technique is to equate a space with the duration it takes to walk a specific distance. This idea is still enshrined in one of the less-common definitions of the English word "mile" found in the *Oxford English Dictionary*, that is, the time it takes to walk a mile. Chaucer uses this notion in describing the divisions marked on the inner ring of an astrolabe—in referring to the long lines that mark groups of five degrees, he writes: "the space bitwene continith a myle wey, and every degree of the bordure conteneth 4 minutes" (1988a [ca. 1391], 664). In Salzman's translation of the Fabric Rolls, he suggests that the "half a space" to which the rolls refer was the time it took to walk half a league (1952, 56). Coincidentally, the circumference of the medieval walls of York is a full league and takes about one modern hour to walk, so it is possible that the "space of time" actually referred to walking half the circuit of the wall, or was even tied to the movement of guards on the wall.

After the break, the workers would work "usque ad horam nonam, et tunc ad prandia sui ibunt" [until the ninth hour, and then they shall go to their dinner]. Here again is a source of confusion, since *nona* literally means the ninth hour and not midday. This ninth hour must not be identified in terms of clock time. It is better conceptualized as the halfway point between solar noon and sunset. Since York Minster was built on an east/west axis with its transepts on a north/south, solar noon would have been easy to reckon—the building could substitute for a compass rose.

The *Ancrene Riwle*, a Middle English guide for anchoresses written in the thirteenth century and widely distributed, says,

"And þe þridde hour after Midday. þat is cleped. hora nona" (1976 [ca. twelfth century], 16). The three women to whom the *Ancrene Riwle* was written probably did not pass through the novitiate in a cloister (Ackerman 1978, 738), and anchoresses were solitary ascetics typically associated with parish churches rather than monastic communities, so their reckoning of time was independent of the daily round of monastic activity driven by St. Benedict's Rule.

The fourteenth century in England was when *none* moved earlier in the day until it came to correspond to the highest point of the Sun, but as Bilfinger demonstrates, authors from the period differ as to whether *none* was at midday or at the third hour after midday although he concludes that at least the English word *none* or "noon" had come to refer to midday by the end of the thirteenth century (1892, 39–42). Moreover, Bilfinger points to evidence of a difference between the vernacular and Latinized ways of indicating time (1892, 60)—he suggests that during this period the "common man" (*gemeine Mann*) viewed midday as *none* and the "educated" (*Gebildete*) knew *none* as three hours later (1892, 60).

As already mentioned, in York's climate, there would be a cognitive problem of determining *none* on cloudy days, regardless of whether it was at the Sun's highest point during the day or at the third hour after or at some time in between. For the construction workers, however, it would have been signaled by a bell—although which bell (York Minster, or St. Mary's Abbey, or some other church/monastery?) and what logic the bell ringer used to determine *none* is not known. The spiritual significance of the period from sext until *none* is based on the Passion story and is noted in liturgical practice (Bilfinger 1892, 64; see Cassian 2000 [fourth—fifth centuries], 59–62; *Ancrene Riwle* 1976 [twelfth century], 16–17; and Aelfric 1875 [ca. tenth century] for examples): Jesus was placed on the cross at sext and died at *none*.

According to the Fabric Rolls, the length of the dinner break varied with the season:

> Post prandium vero a festo Invencionis Sanctae Crucis usque ad festum Beati Petri ad vincula dormire debent infra logium. Et cum vicarii venerint de mensa canonicorum post prandium,

> matister cementarius vel ejus substitutes faciet eos de sompno surgere...
>
> [From the feast of the Invention of the Holy Cross [3 May] to the feast of St. Peter's Chains [1 August], they should sleep in the lodge after dinner. And when the vicars come from the table of the canons after dinner, the master mason, or his deputy, shall get the workers to rise from slumber and work...] (York Minster 1859, 172)

The patron saint of York Minster is St. Peter, so it makes some degree of sense to have the work schedule acknowledge the feast days devoted to St. Peter. The period from *none* to when the vicars would have left the canon hall might have been longer than it took for the vicars to eat their meal. In two English monastic rules, the *Regularis Concordia* (1953 [ca. 970]) and Lanfranc's *Constitutions* (1951 [ca. eleventh century]), there is a period of common prayer after the recitation of the office at *none* and the eating of the meal. The challenge is to relate these rules that applied to monasteries to the routine of the secular canons that served York Minster Cathedral, but if anything, secular canons had a reputation for being more lax in their routines and knowledge of time than their monastic counterparts, as is indicated by comments such as Byrhtferth's: "We know for certain that many city clerics are ignorant of the types of years; but, sustained by the assistance of the fathers—with whose dogs I am unfit to lie down—it is a pleasure to deal with the clerics' idleness" (Byrhtferth 1995 [ca. 1010], 19).

The period of the workers' meal break would have spanned the recitation of the office, the period of prayer after, and the meal at the canon's table—plenty of time to eat and take a nap.

During the winter, the amount of daylight was much shorter, so the dinner break was much shorter as well:

> statim post prandium suum propalam hora competenti sumptum ad opus suum redibunt non expectantes recessum vicariorum de mensa canonicorum
>
> [immediately after the men openly agreeing that an adequate time has been spent on dinner, the men will return to work not awaiting the retiring of the vicars from the canon's table.] (York Minster 1859, 172)

After the dinner break throughout the entire year, work continues “usque ad primam pulsacionem ad vesperas” [until the first bell of vespers]. This bell would announce that the liturgical office of vespers was about to begin. The second bell would be chimed at the beginning of the service, and the third and final bell at the end of vespers. Vespers, itself, was semantically tied to the appearance of the evening star—the term “vespers” comes from the Latin word for evening star, *vespera*. As with the bell of the Blessed Virgin, it is unclear which bell among the many that could be heard would have been the source of the time.

At the first bell, the workers could return to their lodge to drink, and this break lasted until the third bell of vespers. During the summer, they would return to work “quamdiu per lucem diei videre poterunt” [as long as there is enough light of day to be able to see]. During the winter, they would return to work “usque ad pulsacionem companae abbathiae Beatae Mariae quae vocatur le Langebell” [until the ringing of the bell of the abbey of St. Mary’s, which is called the Langebell] (York Minster 1859, 72). This involves an explicit shift in the indicator of time from any of the bells of York Minster to the bell of the nearby St. Mary’s Abbey, which was located just outside the city walls. Bilfinger states that the *Lange Glocke* (or “Long Bell” in English) was used by the authorities to signal curfew (1892, 20).

Cognitively, then, the determination of the time of day, but not of duration, would be tied to the sonic qualities of the environment and observable time cues. These signs are important elements of the medieval timescape, but since they would vary with the season and latitude, they cannot easily be reconciled with the uniformity of clock time. Timescapes involve the interaction of multiple rhythms and include both cultural creations and physical phenomena. As Adam points out, “In the light of their diverse temporalities, ‘nature,’ ‘culture,’ and ‘the environment’ are reconceived” (Adam 1998, 11). Rather than staying conceptually separate, they become phenomenologically intertwined with what Adam describes as the “complex temporalities of contextual being” (1998, 11). The medieval timescape of the construction of York Minster involved considerable cognitive mediation, as well. During the workday, the timescape of the workers overlapped with that which structured the liturgical day. The overlap was only partial, and diverged with the agreement between workers and their supervisors that determined the end of the dinner break in winter and the end of the breakfast break in spring and summer. Importantly, this timescape of the construction workers did not defer to a single device—not a clock, nor a sundial, nor a water clock. The timescape also included at least two belfries—that at York Minster, and that at St. Mary's Abbey.

This document in the Fabric Rolls indicates a very complicated timescape with sonic and visual qualities as well as different seasonal rhythms. Workers and employers formed a community of practice attuned to the sounds of bells, the behavior of the vicars, and either observation of guards walking the walls, or estimations of how long it took to walk a specific distance. Contrary to Thompson and Le Goff, this timescape was sufficient to organize labor and to impose penalties on those workers who did not appear at work on time:

> Item predicti duo magistri cementarii et carpentarius fabricate intererint in qualibet pacacione et ibi notificabunt custodi fabricae et contrarotulatori ejusdem defectus et absencias cementariorum carpentiorum et ceterorum operariorum et secundum moram et absencias cujuslibet deducatur de slario suo tam pro dieta integra quam pro dimidia prout juistum fuerit in hac parte. (York Minster 1859, 172)
>
> [Also the said two master masons and the carpenter of the works shall be present at every pay-day (*pacacione*), and there shall inform the warden and controller of the works of any defaults and absence of masons, carpenters, and other workmen, and according to his lateness (*moram*) or absence deductions shall be made from each man's wages, both for a whole day and a half day, as is reasonable.] (Translation by Salzman 1952, 57)

Based on Lewis Mumford's (1963 [1934]) thesis that the clock was the most important catalyst for the Industrial Revolution, E. P. Thompson (1967) argues that work discipline connected to time arose with the Industrial Revolution; whereas he might have a point with durations of an hour, it is clear from the York Minster rolls that there was a relationship between pay and time—yet not the uniform hours of modern capitalism.

This excursion into the timescape in the community of practice of one particular place and time points out the extent to which the awareness of time was a matter of public perception and part of the web of social practice and relationships. Time was not mediated by individual timepieces but by public signals. Moreover, workers and employers did not look to a single source for indications of time—not even the ringing of the canonical hours by any one of the churches in York was solely used for time reckoning. Multiple time sources, corroboration, and reconciliation between multiple sources were relied upon to determine and probably to dispute the time. There was also a separation of the reckoning of the time of day from the reckoning of elapsed durations. The time of day determined when work started, when breaks took place, and when work ended. This created a seasonally variable rhythm of work. The reckoning of duration, whether it was by "half a space," or by the dinner habits of the vicars, or by an agreed-upon reasonable time to eat, was based on cognitive models quite different from the determination of the time of day—for instance, half a space was the same duration regardless of season. Consequently, unlike moderns who use the clock as a tool for determining time *and* for measuring time, the medieval workers had a different time consciousness that separated these two cognitive tasks. Time was local, fragmented into multiple sources, and required intersubjective corroboration and reconciliation. The need for such intersubjective corroboration persisted well into the period of industrial factory work and was a feature of clocks and watches themselves—the existence of clocks and watches did not imply that they indicated the same time, and Bruegel discusses a labor dispute at the Catskill Furnace and Machine Shop in the 1830s over the employer seeking to install a clock with a bell to determine time in the shop. The issue of contention was that the workers wished to use their own timepieces rather than defer to their employer's (1995, 557). Such distrust of the time kept by the factory is also noted by E. P. Thompson (1967, 86).

Conclusion

The Trinidadian cases and the medieval material indicate complex timescapes in which communities of practice use multiple and contextually embedded sources for determining the time, rather than a single cognitive object for knowing time. Of the multiple sources for time reckoning, some are tied to specific times of day, much like the parrots in Trinidad. Others could be used at any time, for example, the duration of a specific distance walked. Some sources of time information ideally occur at the same moment, thereby allowing corroboration or even synchronization, such as the vespers bell and the appearance of the evening star. They suggest that there are many cues within the environment, some humanly created—for example, the eating habits of vicars, or the soccer-playing habits of Trinidadian youth—which can be used to determine points in time.

Moreover, the medieval material suggests that the cognitive tasks of measuring duration and determining the time of day can be separated. Indeed, before the adoption of mean time, they had to be separate in the European tradition of time reckoning because the length of hours varied with the length of daylight, but there were still tasks that took a set duration regardless of season.

The invention of the clock then combined with the mathematics that allowed the calculation of mean time, which probably occurred by the time of Flamsteed, who was the first Astronomer Royal of Great Britain. Mean time allowed the two cognitive tasks of the measure of duration and the determination of moments in time to be combined in a single tool. But Thompson's argument is circular. If one accepts the Marxist assumption that labor under capitalism is measured by uniform hours, then it is impossible to conceptualize from a Marxist perspective the measure of labor without uniform hours—industrial capitalist labor, by definition, must involve clock time.

The condensation of temporal ideas into clock time has had cognitive consequences, most notably the atrophy of the ability to determine time from environmental cues. As Bergson said about this scientific approach to time, "For a Kepler, or a Galileo, . . . time is not divided objectively in one or another by the matter that fills it. It has no natural articulations" (2005 [1911], 360). Basically, the sophisticated knowledge of a timescape signaled by visual cues and sonic cues diminished in favor of looking at a clock. This is even embedded in the English language. "Clock" comes from *glock*, which is Old English (and German) for "bell," and consequently invokes the sense of hearing; "watch" invokes sight. As Glennie and Thrift (2009) demonstrate, this transformation lagged behind the availability of clocks and watches. Bruegel provides an example from 1830 from Albany, New York, where the local newspaper complains about the two public clocks not being able to be heard throughout the entire city (1995, 553)—if personal clocks and watches were reliable and intersubjectively consistent then there would be no grounds for such a complaint. In contrast, Glennie and Thrift portray a dense signaling of time by bells in Bristol, England, dating to the fifteenth and sixteenth centuries (2002, 162–164).

From a perspective of social function, one could argue, as has been suggested by E. P. Thompson, Le Goff, and even Marx, that the clock was valuable in allowing for the imagination of labor and commerce in terms of uniform durations tied to specific times regardless of season. Yet, from a perspective of thinking about time, the reliance on the clock as a cognitive tool has limited the ability to think about behavioral patterns that vary seasonally. Thompson's essay "Time, Work-Discipline and Industrial Capitalism" (1967) is a useful example. He assumes that the emergence of the clock allowed the organization of labor in time, and that an important shift in time consciousness occurred when laborers used the clock to conceptualize their time in their relationship with their employers. The Fabric Rolls of York Minster demonstrate that previous to the clock there were shared understandings about time between workers and employers, and the context of the 1352 document from York indicates that it was the product of negotiations between workers and their employers. The clock did not usher in a shared consciousness of time, nor did it suddenly allow employers to manage labor in terms of time—both are evidenced by the circumstances in York in 1352.

What makes a difference between premodern and modern systems of telling time is not the device itself, as Thompson indicates, but the combination of the measure of duration with the determination of moments that is adopted by a community of practice. This cultural combination of two cognitive tasks allowed the uniformity achieved in the measure of duration to be applied year-round regardless of the amount of daylight, and to be equally applied at different latitudes. It also required a separation of the cognitive task of reckoning time from the observation of the environment and the Sun, to make it entirely reliant on a device that mediated cognition.

The device then came to shape not only the representation of labor, but also the scholarly study of time. Its embedded concepts became widely used in the construction of many forms of knowledge. As a common tool for conceptualizing time, clocks and watches have constrained concepts of time. The cognitive models embedded in timepieces were developed in a murky past by shadowy figures so that the cognitive mediation provided by clocks is a total obfuscation of time. The radical separation of clocks from nature—as Newton advocated—has resulted in a seemingly arbitrary and shifting relationship between environmental time cues and the cognitive tools used to reckon time.

For many, however, daily tasks do not conform to the clock. The anachronistic rooster is one example of how a creature that was once regarded as wise has come to be viewed as annoying based on the same behavior. Many societies and peoples continue to be influenced by aspects of their environment that have regular rhythms, but are not easily mapped by the clock, and for which clock time provides a maladaptive means of thinking, such as farmers struggling with the milking of cows in relationship to the changing time of markets when there is a switch from standard time to daylight saving time (Prerau 2005, 103–106). As a form of mediated distributed cognition, the clock greatly reduces a cognitive load in some ways, and provides a powerful means of coordinating and representing uniform time across locations and seasons. Yet, it cultivates cognitive impairments in other ways. The challenge in the study of time from the perspective of cultural cognition is to refuse to be limited by the models and algorithms embedded into the clock so that it is possible for other temporalities to be acknowledged and understood.

Chapter 3

Calendrical Uniformity versus Planned Uncanniness

Tollimus autem et abolemus omnino vetus calendarium
[Therefore, we entirely destroy and abolish the old calendar]

—*From* Inter Gravissimus, *Pope Gregory the XIII's bull instituting the Gregorian calendar*

Near two years ago the popish calendar was brought in; (I hope by persons well-affected!) certain it is that the Glastonbury thorn has preserved its inflexibility, and observed its old anniversary. Many thousand spectators visited it on the parliamentary Christmas-day—Not a bud was to be seen!—On the true Nativity it was covered with blossoms. One must be an infidel indeed to spurn at such authority.

—*Horace Walpole (1798 [1753], 160–161)*

Walpole writes of the legendary Glastonbury thorn not cooperating with Great Britain's adoption of the Gregorian calendar in 1752. According to legend, this hawthorn grew from a staff brought to England by Joseph of Arimathea—the owner of the tomb in which Jesus' body was placed. The thorn bloomed on Christmas according to the Julian calendar, and after the calendrical reforms of 1752, it continued to bloom around Julian Christmas rather than the Christmas of the Gregorian calendar. Walpole's wit chides the superstitious and the pope at the same time—the tree refused to cooperate with the pope's decree, and the populace superstitiously believed that mattered.

Environmental cycles do not consistently cooperate with the calendar, which is why many of those who rely on the environment for their livelihood are not content with calendrical time reckoning to determine their activities. In rural Trinidad, the annual cycle of life is palpably influenced by the seasons—dry season, rainy season, Petit Careme (little dry season), and the next rainy season. The harvesting of cocoa, the planting of tubers, and the colors of the forest, such as the bright orange blossoms of the immortelle trees (*Erythrina glauca* Willd.), are all seasonal indicators. At the same time, calendars are ubiquitous. The calendars look like the Gregorian calendars one can find throughout much of the world, but the differences are in the details. Many of Trinidad's national holidays are not tied to a specific day of the Gregorian year—Carnival, Eid, Hosay, Phagwa, and Diwali all shift from one annual cycle to the next (see figure 1.1 in chapter one). In effect, the seeming uniformity of the Gregorian calendar in Trinidad hides a diversity of cycles. In hiding the cycles, it threatens awareness of how cycles interact and intersect—moments of coincidence that can gain social and even spiritual significance. The homogeneous empty time that Anderson (2006 [1983]) sees as an important component in the emergence of the imagination of community among contemporaries within a nation, and which has been identified as a characteristic of secularism (Asad 2003; Taylor 2007), has a consequence of pushing out of consciousness the potential of significance drawn from the interaction of multiple temporalities.

Like the clock, the Gregorian calendar abstracts time from environmental cycles. Also, like the clock, the Gregorian calendar emphasizes uniform duration as the means of reckoning time. Such is not the case with all calendrical systems—particularly those that give equal weight to lunar and solar cycles. Such calendars must confront the difference between the lunar year of around 354 days and the solar year of around 365 days, and develop a means to keep them synchronized—usually by adding an extra lunar month every three or so years. A consequence of such calendars is that "years" are not of equal length.

As with the clock, calendars, as objects of time, shape and direct cognition. Calendars are inscribed condensations of knowledge about cycles. Their being set in stone (sometimes literally) allows for faithful reproduction—a feature of literacy identified by Goody and Watt (1963). In contrast to how Goody and Watt represent the benefits of literacy, written calendars do not automatically allow more information to be retained than oral traditions do. Calendars are organized by logics that choose certain ideas over others, for example, solar years over lunar. So calendars faithfully reproduce their logic, but often at the expense of the reproduction of knowledge about other temporalities. As Elias wrote, "knowledge of calendar time ... is taken for granted to the point where it escapes reflection" (1992, 6), and such lack of reflection easily leads to the unquestioned acceptance of a calendar's logic. Calendars reflect choices about the reckoning of time, and these choices are inscribed in the artifact itself.

The choices made in the design of the Gregorian calendar are to emphasize the solar year, and to adopt the Roman months long since divorced from any natural cycle and their own founding logic. For instance, why is *Dec*ember, the "tenth" month, actually the twelfth? The Gregorian calendar represents time as consisting of empty containers to be filled. This gives the Gregorian calendar an image of flexibility in that all sorts of other cycles can be represented within it. For instance, it is common for Gregorian calendars to indicate the phases of the Moon. Yet, indications of the phases of the Moon receive the same treatment as, say, the dates that payrolls must be submitted. Whereas the seeming flexibility of the Gregorian calendar to incorporate any cycle may seem as a strength, in most instantiations it is devoid of such information—leaving it up to the individual user of the calendar to personalize it, so to speak. As shared artifacts, one could make a case that outside of almanacs, most copies of Gregorian calendars are rather devoid of information.

Noncalendrical knowledge is not to be underestimated and remains important in societies that have adopted the Gregorian calendar. In agricultural areas near forests in Trinidad, the spectacular blooming of the two varieties of immortelle trees are closely tied with seasonal timings. They are eye-catching environmental cues about the time of year. They are also cues that those who engage in slash-and-burn horticulture (primarily the cultivation of tubers) find more useful in timing the felling and burning of trees than calendar time. The blooming of one of the two varieties of immortelle trees signals the coming of the dry season, and the other variety of immortelle blooms during the dry season. The immortelle tree is particularly salient since it is used as a tree to provide shade to cocoa and coffee trees, and, consequently, is ubiquitous throughout the cocoa-growing regions of Trinidad.

There are other well-documented ethnographic cases of environmental cues, as opposed to astronomical cues, being important in marking cycles, such as the month of Milamala for the people of Vakuta in the Trobriand Islands. During this month, there is a swarming of sea-worms (*Eunice viridis*) that is used to synchronize the calendars (Leach 1950, 254; Malinowski 1927, 212–213, 1935, 54; Mondragón 2004). With regard to the swarming of sea-worms around Sumba in Indonesia, Hoskins notes that for different locales on the island, "[t]he amount of agreement between the calendars [of different parts of Sumba] is strongest concerning the moons when the sea-worms are said to swarm" (1997, 345), and she concludes, "I think that both seasonal indicators in the dry season *and* the sea-worms are used to keep the lunar calendar synchronized with the solar year" (1997, 349; emphasis in original), or as one old Kodi man told her, "'[S]tart with the sea-worms. That is where we start ourselves'" (1997, 80). Munn describes how, for Gawans, the seasons are linked with the direction of the prevailing winds, and are used to conceptualize the "wind year" (Munn 1986, 29). Likewise, caboclos (Portuguese-speakers) who live in the Amazon's floodplain in Brazil coordinate their seasonal social and commercial activities not to calendrically reckoned time, but to the changing height of the river, the migrations and presence of certain fauna (fish and birds in particular), and seasonal variations in plants (M. Harris 1998, 75). In classical Greece, animal migrations were used as indicators of seasonal changes, such as in Hesiod's *Works and Days*: "Take heed when you hear the voice of the crane from high in the clouds, making its annual clamour; it brings the signal for ploughing, and indicates the season of winter rains" (1998 [ca. eighth century BC], 50). The reckoning of time through noncalendric and nonastronomic means is thus widespread.

Noncalendric logics tend to emphasize the observation of phenomena from which it is possible to count units of time, for example, the Trobriand practice of counting lunations from the spawning of the sea-worm on the sea's surface (Malinowski 1927, 212). Moreover, counting is a useful hedge against not being able to observe particular phenomena. For instance, if one has charted the phases of the Moon, it is possible to know what phase it is on a cloudy night based on the number of days that had passed since it was last observed—possibly this was one of the cognitive functions of the Paleolithic artifacts Marshack discusses that seem to chart lunar phases (1972, 43–55).

In another example, Stephen, who visited the Hopi in the 1890s, provides a detailed account of the use of the environment to allow a countdown to the ceremonies performed on the winter solstice:

> 1892
>
> Tuesday, December 6
>
> Kwa'chakwa, Sun chief (Ta'wa moñwĭ), still continues his observations at sunset, accompanied by Su'pelă, chief of Winter solstice ceremony, on the flat house roof of the house group I live in... He says that in three days the sun will set in the notch made where Eldon Mesa intersects.
>
> Saturday, December 10
>
> Just as the sun began to peer over the eastern plateaus, Ho'ñi, standing on the roof of his maternal family house, the house of Sha'lĭko, announced the Winter solstice ceremony. Men are to meet in all the kivas on the fourth day from now.

Then several days later, Stephen notes:

> Saturday, December 17, IV day
>
> The evenings have been too cloudy and squally since the tenth to admit of marking place of sunset, but Kwa'chakwa has not concerned himself to watch the sunset, nor has anyone else given heed to the sunset since Kwa'chakwa set the time (on the sixth) when announcement should be made by Ho'ñi on the tenth. (1969 [1936], 39)

This description of Hopi time reckoning combines an observation of the Sun's position on the horizon and counting (Stephen 1969 [1936], 1041). The winter solstice is actually not sighted, nor do the Hopi seem concerned with the cloudiness that obscures their vision. The point is to anticipate the day of the solstice, and consequently, once there is a cognitive means of predicting when the solstice will be, there is no longer a need to actually observe the solstice—instead, one can simply count. This behavior seen among the Hopi might be a widespread pattern among societies that watch the skies—as Brennan has written of the winter solstice at Newgrange in Ireland, a megalithic mound in which the Sun's light is directed down a long passage into a chamber on the winter solstice and adjacent days, the fact that the Sun shines into the chamber for several days before the solstice and after the solstice allows the anticipation of the solstice and mitigates against the problem posed by the possibility of cloudiness on the solstice itself (1983, 43).

There is ethnographic support for societies in which counting is more important than an agreed-upon calendrical system. Turton and Ruggles describe how Mursi, a pastoral group in Ethiopia, closely tie the annual cycle of their activities, known as *bergu*, to the seasons: "Any Mursi over the age of about 12 is able to recite a list of seasonal activities which are associated with these numbered subdivisions of the *bergu*" (1978, 587). This common cultural knowledge does not produce agreement in time reckoning, much less a shared calendar. Instead, Mursi have an "agreement to disagree about the number of the *bergu*" and this disagreement is "an institutionalized part of the system" (1978, 589). The way in which the timing of significant events is determined is through the estimation of lapsed time between one event and another (1978, 593). In his survey of Tiv concepts of time, one of Bohannan's conclusions with regard to all the phenomena used to reckon time—solar and lunar cycles, dry seasons, and markets—"*counting* is of special importance" (1953, 257; emphasis in original).

Sometimes calendrical logics based on counting are not clearly tied to any astronomical referent. For the Maya, one of the basic units of time, the 260-day cycle called the *tzolk'in*, has come down to us as a count that is not explicitly linked to any celestial cycles. This lack of an obvious astronomical referent has led archaeologists and ethnoastronomers to speculate on its origin and to debate whether it was tied to some astronomical cycle such as that of Venus (Justeson 1989), or to an agricultural cycle (Rice 2007, 35–36). Yet, one could imagine future scholars being stumped by the months in the Gregorian calendar. In the case of both these calendars, their origins recede into the background of cultural amnesia. The logic of the calendar takes on a significance all its own, and does not need an original function or cause.

The problem facing those trying to understand any counting-based calendrical logic is that self-referentiality trumps any obscure origin. As with clocks, such self-referentiality puts the calendar beyond easy dispute, and places the underlying logic of the calendar outside of the consciousness of most of its users, including scholars. It should be mentioned that such self-referentiality is not limited to calendars, but is manifest in many objects that are designed to make the results of highly technical knowledge and calculations available to their users in a simple representational form. The simplicity of the representation relieves users of the need to master the technical knowledge.

Civilization, Calendars, and Artifactually Mediated Astronomy

With calendars, there has not emerged a narrative of transformation from precalendrical to calendrical timekeeping that parallels the narrative of the shift from preclock to clock-based time. In contrast to clocks, calendars are frequently associated with "ancient civilizations," even going so far as to be a sign that an ancient society is a civilization. As chapter two pointed out, the study of clocks divides the history of time reckoning into clock-based and pre-clock thought, with the suggestion that before clocks time reckoning was "irregular" and "close to nature"; in studies of the evolution of calendars, there are no such statements, but instead there are discussions of how the problem of reconciling astronomical cycles—particularly lunar and solar—has produced culturally variable results (see Aveni 2002; O'Neil 1975). The field of ethnoastronomy often seems as much about the construction of calendars as about the cultural creation of astronomical knowledge—the two are closely tied. So in contrast to clocks, calendars are not enmeshed in the emergence of industrial capitalism or globalization. The closest association of calendric time with modernity is Anderson's argument that one of the developments that fostered a sense of imagined community was newspapers' displays of dates to create senses of "calendrical coincidence" (2006 [1983], 33).

Even though astronomical cycles are at the center of much of the study of ancient calendars, some of the most studied cases of calendars—the Julian calendar, the Athenian civil calendar, the Mayan calendar, and Christianity's Easter tables—are cognitive tools to reduce reliance on direct astronomical observation. Jones's extensive study (1943) of the debate over the Easter cycle shows that its end result was to create tables that relieved the need to rely on astronomy. As Jones says of these tables: "It must be remembered that in the poorest ecclesiastical centres, books of history were not deemed necessary, but an Easter-table was essential" (1943, 116). It is the self-referentiality of the cognitive tool that relieves the cognitive burden for its users. This allows the widespread cultural sharing of the representation produced by the tool as well as the logic embedded in the tool that produces the representation. This wide distribution of representational form and the logics behind it occurs even when the users are unaware of those logics.

Our knowledge of prehistoric ethnoastronomy derives from the study of artifacts and how structures and sites align with celestial events, but this too easily leads to an interpretation of this material record as having an observational function. In his study of Roman devices for reckoning astronomical and meteorological cycles, Lehoux suggests that the link made between the rising of certain stars and weather patterns was for purposes of prediction (Lehoux 2007, 8–9, 12). In general, it is a greater power and of greater use to be able to know ahead of time when the solstice, say, will occur than to be able to merely note that it has occurred; this power is noted in Greek, Chinese, and Mesoamerican systems, but typically is not applied to megalithic sites. In fact, one of the most remarkable examples of technology from classical Greek society is the Antikythera Mechanism found by a Greek sponge-diver in the Mediterranean. This mechanism, which consists of gears and wheels, charts the cycles of several celestial objects (de Solla Price 1974). One of its abilities, since it charted the intersection of lunar and solar cycles, would have been the prediction of lunar and solar eclipses (Hannah 2008, 31). Consequently, its power was not merely to guide observation on any given day, but to predict significant celestial alignments.

Some of the oldest tools of time reckoning seem to have the ability to predict and chart cycles. The phases of the Moon carved into a bone dated to 20,000 BP from France is quite possibly the oldest calendric tool known (Marshack 1972, 43–55; Aveni 2002, 58), and while it could have been used to guide observation of the Moon, it could have just as easily been used to know what the phase of the Moon was when one could not see the Moon due to clouds. In ancient Rome, parapegmata were important tools "to keep track of the increasing number of significant phases," such as the occurrence of specific weather with stellar, lunar, solar, and even mercantile cycles and phases (Lehoux 2007, 12). Parapegmata came in two artifactual forms. One was inscriptional, with holes for pegs placed next to labels of the important phases in each cycle. These were used by moving the pegs from one hole to the next. The other form was literary—written tables or calendars in which phases of cycles were recorded. In classical Greek and Roman societies, parapegmata served as cognitive tools in which cycles were represented by the artifact so that one could keep track of time by moving the peg rather than by observing the skies—a useful ability in cloudy weather (Lehoux 2004, 237–243, 2007, 55–64; Taub 2003, 24).

Often, parapegmata charted multiple cycles and thereby captured how these cycles intersected and interacted. About Roman parapegmata, Lehoux writes, "Time is just intercyclical, and Roman parapegmata need to be understood in this light... Latin inscriptional parapegmata can be seen to be tracking, not so much several independent cycles at once, but more the ways in which the difference cycles *interact*" (2007, 29; emphasis in original).

Literary parapegmata provide some of the clearest evidence of cognitive tools that were calendrical but not observational. When Ovid wrote about the *nones* of January, in other words, the fifth, he indicated the position of the constellation of the Lyre through reference to rain: "Signs that the Nones are here will be dark clouds, pouring rains, and the rising Lyre" (Ovid 1995 [ca. 8–14], book I, line 315). Since one cannot view constellations under conditions of dark clouds and pouring rains, Ovid's insight allows readers to know the time and the position of the Lyre despite not being able to observe that constellation.

The focus on the interaction of cycles makes parapegmata, the Antikythera Mechanism, Mesoamerican calendrical systems, and likely many ancient ethnoastronomical technologies different from the Gregorian calendar. The former embrace temporal multiplicity and draw significance from representing the intersection of cycles; the latter embraces temporal uniformity and subsumes all cycles into a unified grid of time with relatively homogeneous units. What are the cognitive and even ideological consequences of this difference?

The Planned Uncanny

In an approach to calendars that seems to fit Mayan calendars and classical Greek and Roman parapegmata better than the Gregorian calendar, Henri Hubert argued that "[f]or religion and magic, the object of a calendar is not to measure time, but to endow it with rhythm" (1999 [1909], 49), and added, "*the successive parts of time are not homogeneous*" (1999 [1909], 50; emphasis in original). As Munn has pointed out, Hubert emphasizes the qualitative aspect of time (1992, 95). Instead of the uniform time typical of the Gregorian calendar, what is important to Hubert are critical dates and the relationship between these dates. He concludes, "[T]he institution of the calendar has not as its sole nor doubtless its main object the measurement of the flow of time, considered as a quantity. It originates not in the idea of a purely quantitative time, but in that of qualitative time, composed of discontinuous parts, heterogeneous and ceaselessly revolving" (1999 [1909], 81).

"Person, Time, and Conduct in Bali" by Clifford Geertz provides an example of how calendars can function in ways different from those embedded in the Gregorian calendar. He describes the Balinese calendar as not emphasizing the measure of duration, but instead seeking to classify time "into bounded units not in order to count and total them but to describe and characterize them, to formulate their differential social, intellectual, and religious significance" (1973, 391). He describes the Balinese calendar as "clearly not durational but punctual" (1973, 393). Because the calendar involves the interaction of multiple cycles, the occurrence of festivals is "spasmodic" (1973, 394) and life is "irregularly punctuated by frequent holidays" (1973, 395). In this regard, I think Geertz conflates uniformity with regularity—the occurrence of holidays in Bali is rhythmic, but the rhythm is not one of monotonous uniform beats.

Repetition and periodicity are important features of many ritual and religious traditions and are components of what Gell identifies as the "ritual manipulation of time" (1992, 37). Whereas such ritual manipulation has often been viewed in terms of creating a liminal time of symbolic inversions (Leach 1961), what is at stake in these contexts is a dialectic of sequence and simultaneity (Gell 1975, 1992, 52–53). In such approaches, what gives the time significance are the rituals that are performed, whereas Hubert and Durkheim's suggestion is that what gives the rituals significance is the time at which they are performed—a point that has received far less attention from ethnographers than symbolic inversion has.

Such congruence of time keeping and ritual is well documented for the Maya. Rice (2004) describes how the *may ku*, the city designated as the political and religious center of a region, was determined by cycles in Mayan time reckoning, such that the designation of a city changed with each *may*—a 256-year cycle. A *may* that marked the recurring congruence of several cycles. For the Maya, the numbers 13 and 20 were of great importance in time reckoning—a *winal* consisted of 20 days, an approximate solar year (*tun*) of 18 *winals* plus five auspicious days, a *k'atun* of 20 *tuns*, and a *may* of 13 *k'atuns* (Rice 2004, 59). The calendric year of 13 *winals* was distinct from the solar year. If one takes the number of days in 13 *winals* (260) and multiplies it by the number of days in a *tun* (365), one gets 93,600 days—which is around 256 solar years and equivalent to the *may* cycle. So a *may* represented the cycle of recurrence of the interaction between the Maya's two different forms of calendrical time reckoning. The units of these calendars "had primary importance in prognostication and in tracking mythoritual time" (Rice 2004, 59). Puleston goes even further to suggest that the social disintegration associated with the beginning/end of the cycle of thirteen *k'atun* "became a self-validating myth" (1979, 70).

Such complicated calendrical logics were not only of significance in determining the religious and political center of the Maya, but also for charting a variety of astronomical events—"Postclassic astronomical tables provide temporal information concerning the timing of celestial events, showing that the elite could anticipate coming astronomical events; this would aid preparations for activities of their own selection in the guise of submission to the divine will" (Justeson 1989, 104).

Hubert's claim that religious calendars emphasize rhythm and timing can be augmented by the suggestion that they also cultivate religious sensibilities through the ability of these rhythms to, in a sense, allow the anticipation of uncanny coincidences. In his book *The Idea of the Holy*, Rudolf Otto (1950 [1917]) suggests that an important phenomenological component of religious experience is the sense of mystery, what he calls "mysterium." This is a sense of something "that which is hidden and esoteric, that which is beyond conception or understanding, extraordinary and unfamiliar" (1950 [1917], 13). Crucial to the development of this sense is a sense of awe or dread (1950 [1917], 16), a sense of power (1950 [1917], 20), and a sense of urgency or energy (1950 [1917], 23), all of which are tied to something "objective and outside the self" (1950 [1917], 11).

Freud addresses such a sense in his essay "The 'Uncanny,'" As I have argued elsewhere (Birth 2011a), while Freud interprets uncanny experiences as a problem of reality testing and repetition compulsion, his claim that his superior rationality and reality testing allowed him to dismiss the uncanny exists in tension with his admission of a certain reality to the uncanny: "As soon as something *actually happens* in our lives which seems to confirm the old, discarded beliefs we get a feeling of the uncanny" (1976 [1919], 639; emphasis in original). Even though Freud assumes that the uncanny is an illusion that is grounded in the repetition of some unconscious anxiety, he is unsettled by it. The phrase "actually happens" haunts Freud enough for him emphasize it.

With regard to timing, it is possible to explore the link between coincidences that, in Freud's terms, "actually happen," Hubert's interest in sacred rhythms, and Otto's arguments about the foundations of religious experience. Social rhythms that foster the collective recognition of uncanny coincidences can inspire religious awe. Durkheim's discussion of the corroboree emphasizes its timing. He points out, "The life of Australian societies alternates between two different phases" (2001 [1912], 162). About the first phase, when families are dispersed, he says, "The dispersed nature of the society makes life rather monotonous, lazy and dull" (2001 [1912], 162). But the other phase of the year, when the corroboree takes place, is quite different: "Once the individuals are assembled, their proximity generates a kind of electricity that quickly transports them to an extraordinary degree of exaltation" (2001 [1912], 162). This is the context in which "collective effervescence" forms the foundation of conscience and the separation of the sacred from the profane (2001 [1912], 164). He concludes, "Therefore it is these effervescent social settings, and from this very effervescence, that the religious idea seems to be born. And this origin seems confirmed by the fact that in Australia, strictly religious activity is almost entirely concentrated in the times when these assemblies are held" (2001 [1912], 164), and he finally adds, "By gathering together almost always at fixed times, collective life could indeed achieve its maximum intensity and efficacy, and so give man a more vivid sense of his dual existence and his dual nature" (2001 [1912], 164–165). In effect, social rhythm can produce experiences akin to Otto's mysterium.

For Durkheim, the corroboree is a regular, predictable event. Freud also allows for the uncanny to be predictable. This is implied by Freud's use of a work of fiction, Hoffman's "The Sand-Man." If, as Nietzsche argues in *The Birth of Tragedy* (1956), the plot of tragedy is known and this knowledge gives its affective import to the audience, then the use of a work of fiction to demonstrate and explore the uncanny suggests the possibility that the uncanny is expected.

Predictable, significant coincidences are potentially uncanny, then. Through the cultivation of predictable coincidences, some traditions encourage a sense and experience of mystery (Birth 2011a). The cultivation of predictable coincidences highlights the coexistence of independent cycles and encourages attention to the moments they intersect. The dominant logic of uniformity in the Gregorian calendar has made that calendar ill suited for generating uncanny timings. Indeed, many holidays are not produced by coincidental timings, but simply are assigned a date by the state or a religious body. Even holidays that are the product of coincidental timings, such as Easter, exist in the consciousness for the users of the calendar as a date associated with the season of spring, not as the intersection of lunar, solar, and weekly cycles.

In Trinidad, because of the consciousness of different calendars, the seasonal timing and coincidence of ritual observances of different religious traditions is a source of the uncanny experience of something outside the self. The Hindu celebration of Diwali and the Roman Catholic celebrations of All Saints' Night and All Souls' Night nearly coincide. All three involve the symbolism of light, with Hindus lighting small clay lamps with cotton wicks on Diwali; Roman Catholics lighting candles outside the home on All Saints' Night; and people of all faiths gathering to light candles on the graves of loved ones on All Souls' Night.

The first time I witnessed the celebration of All Souls' Night, those at the cemetery highlighted the parallels between Diwali and All Saints'—to them both represented the victory of light over darkness. About Diwali, a Hindu man told me:

> Diwali has to do with lightness over darkness. Besides natural darkness, which is what we have at night, there is also human darkness. Human darkness is when you do not light your heart...There is a material way about Diwali. We light material lights, and these lights make people happy...Material light and light in the heart are tied together.

On All Souls' Night, candles are lit at the graves of the deceased. As people searched for the graves of loved ones, one Creole man said to me: "The candles are for darkness being changed to light."

Diwali is one of the most important celebrations in the Hindu year. It consists of a puja—a ritual offering—to the goddess Lakshmi. Lakshmi is a benevolent goddess who blesses marriages and homes, and is associated with domestic tranquility and prosperity. In addition to the special puja to Lakshmi, Diwali celebrations include the lighting of *deyas* (small oil lamps), after the puja both at the Hindu mandir (temple) and at people's homes. The homes are then thrown open to visitors—anyone who visits is offered food. Many non-Hindus visit their Hindu neighbors on this night and are treated to heaped plates of Indian food.

In Anamat, my field site in Trinidad, there is also a community puja for Diwali that usually occurs a few days beforehand. This puja is accompanied by a pageant that features local talent performing bhajans and traditional dance, and often a reenactment of a portion of the Ramayana. Not only do many non-Hindus attend the puja and enjoy the following show, but in both years in which I was in the audience, non-Hindus played roles in the show itself—either playing bit parts in the reenactment of the Ramayana, or playing in the musical ensemble that accompanied the singers.

All Saints' Night is a Roman Catholic commemoration of the saints of the church, and is celebrated with the lighting of candles at the threshold of one's property. All Souls' Night commemorates deceased loved ones, and is celebrated with the lighting of candles in the cemetery over the graves.

Just as Diwali attracts non-Hindu participants, All Souls' Night also is a community event. Many Hindus and Muslims go to the cemetery on that night. I remember one woman, whom I shall call Tantie. On All Souls' Night, she sought help in locating the grave of her late husband in order to place a candle on it. There was some difficulty in this. Tantie was a Muslim, and the funeral for her Hindu husband had not been a Hindu cremation, but a Muslim burial. The Muslim tradition is to have a plain burial with no grave marker. For Tantie to locate her husband's grave required finding somebody who remembered the burial and could locate the approximate location in relationship to the main landmark in the cemetery—a large tree near the summit of the hill.

Recently, many Hindus have opted for cremation at the Caroni River, but since cremations have only been allowed since 1953, there are still Hindus who go to the cemetery on All Souls' Night because they have family members buried there. Still other Hindus go for the same reason that many non-Hindus attend the Diwali puja and pageant at the Hindu mandir: it is a community event of interest and aesthetic beauty. The common theme of the victory of light over dark creates a momentary bond and commonality that seems to attract people to witness its manifestation—it was about this time of year that an elderly Hindu man said to me, "Every religion supports one another."

This ecumenical coincidence is part of the Trinidadian timescape. How might the coincidence of such different religious traditions be explained? The Trinidadian cultural response is to engage in a form of creolized ecumenical theology—that the similarities between religions reveal a deeper truth than do any of the religions individually.

Tantie always emphasized the coincidence of these holidays as proof of there being a single God behind religious diversity. She identified herself as Muslim, but would resort to Catholic novenas to ask for divine help in dealing with the struggles of daily life. She carefully maintained the small Hindu shrine on her property that had been built by her husband, a Hindu holy man. Across the street from her lived Lal and his family. Lal came from a prominent Hindu Indian family in the village, but due to family squabbles, did not inherit any land. He made his living as he could, and worked for a nearby Christian school. Lal was related to the founders of the local Baptist church, and allowed them to hold Bible studies underneath his house. I have been to Bible studies there where, while one of the Baptist pastors was preaching, one of Lal's daughters quietly got up to light a *deya*, at twilight, as part of the family's expression of devotion to the Hindu gods and goddesses. Lal's wife, Cynthia, was very vocal about the importance of religion in living a good life, but also about the problems caused by some people using religion to divide the community. Like her neighbor Tantie, Cynthia pointed to her perception of commonalities between the different faiths manifest in Anamat. The intertwining of practices from different religious faiths primes such Trinidadians to experience the coincidence of Diwali and All Saints' as predictable uncanniness.

Uncanniness on a Global Scale

The cemetery is a place of coincidences—a place where the timing of Hindu and Catholic traditions coincide, and a place, by virtue of the celebration of the dead and the time of the year, that evoked my memories of the Halloween. According to Hindu tradition, this is the darkest night of the year, and darkness is associated with evil and death. This claim resonates with the celebration of Halloween derived from the Celtic Samhain. Samhain has been associated with the dead, and reputedly also involved lighting a large fire (Dalton 1972, 224), although the lack of primary sources about the holiday make it difficult to infer anything about its celebration with certainty (Hutton 1996, 360–370). Moreover, the little evidence that exists for Samhain makes no reference to celestial events, even though Celtic religion reputedly emphasized celestial timings for its holidays. For Samhain, one is simply left to surmise what its timing and significance might have been through the lens of its placement in the Julian calendar, with that calendar's uniformity, rather than in a Celtic context. In any case, there is enough evidence about Samhain's timing and significance to include it in the mix of uncanny coincidences of religious commemoration.

It is useful to move away from a local, Trinidadian interpretation of this particular time of year to consider the matter more globally. About the celebration of All Souls' Day in Guinea-Bissau, Brooks emphasizes the resonances between Portuguese Catholic practice and African celebrations, noting that "All Souls' Day on November 2 and All Saints' Day on November 1 'fortuitously' coincided with the time African societies celebrated harvest time and the beginning of the new year" (1984, 2). Brooks attributes this coincidence to pre-Christian practices in Portugal that involved "a pagan festival of death following harvest time and the onset of winter." Brooks associates this festival with the antecedents of modern Halloween.

In effect, there is a congruence of pre-Christian European celebrations, Hinduism's Diwali, and practices in the Guinea-Bissau region of Africa. This is the sort of coincidence that hearkens back to the old debate in anthropology over diffusion versus independent invention (see G. Elliott Smith et al. 1927), although as modern anthropology has drifted away from an ethnological and comparative focus, this problem is often less easily recognized, and is dealt with, as Brooks does, from a local perspective.

Hindus offer an ethnoastronomical explanation for Diwali. It is a religion that is interested in astronomical phenomena in ways that no longer engage Christianity. In particular, Hinduism emphasizes the coincidence of celestial bodies as crucial to understanding one's fate and making wise decisions. Diwali is said to fall on the darkest night of the year based on the decrease in sunlight after the fall equinox, and a new (or dark) moon after what, in the West, is known as the harvest moon—a full moon that rises soon after sunset and is reputed to be brighter than other full moons.

Christians once paid attention to such astronomical cycles, and these cycles gave significance to some holidays and determined the dates of other commemorations. In an essay attributed to Anatolius of Laodicea in the third century, the timing of Easter was described in terms of ethnoastronomical significance:

> as long as equality between light and darkness endures, and is not diminished by the light, it is shown that the Paschal festival is not to be celebrated. Accordingly, it is enjoined that that festival be kept after the equinox, because the moon of the fourteenth, if before the equinox or at the equinox, does not fill the whole night. (1926 [third century], 149)

From the period of Charlemagne, we have an example of Alcuin writing to Bishop Arno of Salzburg to encourage the celebration of All Saints' Day as a replacement for pagan rituals in which the symbolism of light is highlighted:

> Kalendis Novembris solemnitas omnium sanctorum...Quod ut fieri digne posit a nobis, lumen verum, quod inluminat omnem hominem, Christus Ieusus inluminet corda nostra, et pax Dei, quae exsuperate omnem sensum, per intercessionem omnium sanctorum eius, custodiat ea usque in diem aeternitatis.
>
> [With regard to the solemnity of All Saints' on the first of November...So that the appropriate thing may be done by us, may the true light, Jesus Christ, because it illuminates all people, illuminate our hearts, and the peace of God, which surpasses all understanding, through the intercession of all of the saints, keep peace until the day of eternity.] (Alcuin 1994 [eighth century], 321)

McCluskey argues that "Alcuin juxtaposed Christ, the true light, to the pagan's feast's theme of death at a season of growing darkness" (1998, 64). Yet, Alcuin was not addressing Diwali or harvest celebrations in Guinea-Bissau, and these celebrations occur at the same time of year, as well. There is an uncanny coincidence of celebrations at this particular time of year.

In 1733, when there were still multiple calendars in Europe, Isaac Newton also suggested a tie between astronomy and religious commemoration in Christianity:

> They who began first to celebrate them [Christian holidays], placed them in the cardinal periods of the year; as the annunciation of the Virgin Mary, on the 25th of March, which when Julius Caesar corrected the Calendar was the vernal Equinox; the feast of John Baptist on the 24th of June, which was the summer Solstice; the feast of St. Michael on September 29, which was the autumnal Equinox; and the birth of Christ on the winter Solstice, Decemb. 25... (1998 [1733], 143)

Easter is tied to Passover, and Passover is determined by astronomical events. But why is it that these events occur around the spring equinox when light is ascending? Both the full moon and the vernal equinox are observable.

The resonances between Hinduism and Catholicism in Trinidad and with decontextualized samples about the sacred dimensions of celestial events from Guinea-Bissau, the Celts, Anatolius, Alcuin, and Newton stirs a sense of uncanny similarity between societies that are not directly connected. If every day in the secularized Gregorian calendar is the same, it is hard to grasp why certain periods have so many sacred coincidences.

It would be easy to attribute all of this to coincidence, or to fall back on the theological interpretation of Tantie and Cynthia that the similar timing of these events is because each tradition really deals with the same God—but I am not theologically comfortable grouping together Hinduism, Catholicism, and pre-Christian European traditions. Still, the uncanny coincidence disrupts the uniformity of the Gregorian calendar as a cognitive artifact. As I have done work on time, I have come to think that there is something more—that Hindu, Catholic, and the European traditions that determine the timing of Diwali, All Saints', All Souls', and Halloween are referencing something observable now obscured by the use of the calendar. I have come to think that there is something special in terms of time about this period of the year, and that this special dimension is related to cycles of the Moon and the Sun—in effect, that different religious traditions that have had a history of sophisticated astronomical observation to time religious observances have all noted the same phenomena and these phenomena then become the basis for religious interpretation of the sacredness of the time. But since these phenomena are mediated by calendars as cognitive artifacts, lay awareness of the astronomy behind the calendar has diminished or even disappeared to be replaced by coincidences between different calendars that foster ecumenical musings. Referring to observable celestial events reveals an ethnoastronomical logic hidden by the calendar's cognitive mediation. It shifts the coincidence from religious practice in uniform time to the intersection of lunar and solar cycles.

As I argue elsewhere, "Societies differ in whether they cultivate planned experiences of the uncanny, and how societies accomplish this. In fact, even though it is obvious it bears to be stated: a coincidence between the practices of two societies does not mean that there must be a coincidence between the practices of all societies for a sense of the uncanny to emerge" (2011a, 130). This can be seen in figure 1.1 in chapter one. In this figure, it can be clearly seen how the Islamic holidays move in relationship to Christian and Hindu holidays. Even so, many Trinidadian Muslims note the coincidence of the Christian and Hindu celebrations, and eventually the Muslim holy days coincide with Hindu and Christian celebrations in particularly auspicious timings.

The interaction of cycles, then, can involve the creation of the uncanny by reference to features of the environment that "actually happen" (Birth 2011a). The coincidence of Diwali and All Saints' is part of the annual cycle of holidays. The uncanniness derives from the coincidence of two holidays from two different religions representing very different cultural traditions within Trinidad. Some cognitive tools, like parapegmata, almanacs, or horoscopes, are well documented as having that potential. Indeed, astrology emphasizes the significance of celestial coincidences. But the representation of these holidays as dates and not as astronomical coincidences in the material manifestation of the Gregorian calendar so completely mediates cognition that most users of the calendar as a cognitive artifact do not know the original astronomical logics and algorithms that determined the holidays—how many Christians can articulate the rules that guide the determination of Easter?

So the cognitive power of calendars that encourage temporal uniformity includes the power to create what seem to be coincidences between different cultural traditions when these coincidences are based on observations of celestial events. The coincidence of Diwali and All Saints' is a consequence of calendrical uniformity, and allows the relationship between these two holidays to foster a sense of mystical ecumenicalism beyond the specific religions involved. This, in fact, is diametrically opposed to Alcuin's purpose of juxtaposing All Saints' to pagan holidays in order to compete with them. The uniformity of the calendar, then, creates a sense of deep religious uniformity rather than revealing competition over what religious significance should be attached to a particular new moon.

Strategies of Calendrical Uncanniness

Astronomical and seasonal uncanniness are always local because the cycles that indicate them are local, not global. Consequently, cognitive artifacts that chart cycles that intersect to evoke a sense of uncanny timings are also locally calibrated. Globalization and diasporic populations therefore pose a significant challenge to how a calendar makes temporal coincidences uncanny. There have been three main strategies for reconciling calendrical systems that span large areas of the globe with the local embeddedness of astronomical and seasonal cycles. Each strategy involves different cultural choices with different types of cognitive artifacts.

The already-mentioned Hindu calendar has one solution—since astronomical cycles are local, the Hindu calendar has a global structure that is locally adapted. All of the important events in life must be charted astrologically, and certain events, such as weddings and important business transactions, are performed on auspicious days. The result of this calendric system is that it diffuses the power of coordinating events in time to local specialists who consult specialized books and tables. The emphasis on astronomic alignments and specialized knowledge cultivates a sense of good and bad timing among Hindus, and is quite distinct from a view of time as continuous, uniform, and homogeneous. Some of the specialized knowledge is represented in calendric form—in Trinidad through Gregorian calendars in which auspicious days are indicated.

A very different solution is the Jewish calendar. This is a combined lunar and solar calendar in which the relationship of lunar phases to the solar year is important in determining holidays. Sacha Stern writes, "The proper timing of rituals is a central theme of mishnaic and subsequent rabbinic law, arguably to the point of an obsession" (2003, 47). Many of the holidays are not only linked to specific phases of the Moon, but also are part of the agricultural cycle of Israel. As a result, the Jewish calendar synchronizes the celebration of the faithful throughout the world to the local cycles of Israel. For instance, synagogues throughout the world announce the precise time of the beginning of the new moon in Jerusalem (Rosen and Rosen 2000, 265). This sort of calendar is built on the assumption of a common identity established through a common practice associated with ancient cycles in the homeland. The sense of mystery that Otto describes becomes combined with a profound sense of heritage. In charting the emergence of the Jewish calendar, Stern strongly argues in favor of its raison d'être being the creation of "a single, identical calendar in all the Palestinian and Diaspora communities" (S. Stern 2003, 232) and goes on to suggest that this was done in explicit contrast to the calendrical diversity of the place and period in which this calendar emerged (S. Stern 2003, 241). First, fixed calendric rules emerged in the Amoraic period (third–fifth centuries AD), with a fixed calendar emerging by the time of the writing of the Palestinian Talmud in the fourth century AD (S. Stern 2003, 255).

The Gregorian calendar originated as a church calendar, but has become the dominant secular calendar throughout the world. The calendar, then, is a sort of palimpsest in which national and secular celebrations get encoded into what was originally a Christian calendar (Birth forthcoming). The relationship of this calendar to particular locales is a combination of international agreements on time standards and national politics. The day begins at midnight on the international date line, and in that sense the calendar is tied to a specific meridian, but the countries that are near the date line choose a side—their day does not begin so much because of the date line but because of political decisions about what their time should be in relationship to the date line. This is the only variation in the calendar. The beginning of the new year only varies by hours, not days or months, and most major holidays are scheduled with specific calendrical dates rather than being tied to environmental or astronomical events. Indeed, there are many holidays that are defined solely by calendrical dates and nothing environmental or astronomical, for example, Cinco de Mayo, May Day, and the Fourth of July.

This makes the Gregorian calendar different from the Hindu and Jewish calendars. Since the calendar is global and abstracted from astronomical and environmental cues, it does not become much of a source of identity or experiences of the uncanny tied to place or timing. Even seasonality is not represented in this calendar—after all when the Summer Olympics were held in Sydney, they occurred during the late winter and early spring in Australia. The calendar becomes a set of temporal units that the state can decide to fill, and holidays become matters of political will and action rather than a reflection of experience. To some extent, this was true of the Roman Empire. January 1 was a state-sanctioned beginning of the year. As Feeney says of this date, "Why, then, did the Romans begin the year in the dead of winter? The Romans themselves were not sure why their civil year began in January" (2007, 204); and, for that matter, the naming of months after emperors is also a practice of state power defining time rather than the intersection of disparate cycles. Consequently, I disagree with Anderson's (2006 [1983]) assessment that homogeneous empty time emerged as a result of print media, but instead see homogeneous empty time as a necessary step in governmental appropriation of time that predates print media (Birth forthcoming). Print media were merely useful tools for a practice that had already existed for almost two millennia.

It must be remembered that the origin and history of the Gregorian calendar is tied to imperialism. With regard to this calendar's predecessor, the Julian calendar, when Julius Caesar instituted his calendrical reforms, he, in effect, stripped local priests and astronomers of power. When Pope Gregory the XIII (2002 [1582]) decreed a new calendar and calendrical logic in his papal bull *Inter Gravissimas* [Among the Most Serious of Matters], it was a symptom of the Counter-Reformation and an effort to unite not merely the Church but all Christendom. When the non-Catholic states in Europe adopted Pope Gregory's calendar it was not out of deference to the pope but out of the desire to create a uniform time for trade and colonial administration (M. Smith 1998). Whenever countries with non-Christian populations adopt the Gregorian calendar it is a symptom of European hegemony. It is very easy to find Gregorian calendars in which Muslim, Hindu, or Jewish holidays are represented. In contrast, it is very difficult to find a Muslim, Hindu, or Jewish calendar in which Christian holidays can be found. Hoskins notes, "[I]t is increasingly difficult to find groups whose original, preliterate temporalities have not been clouded or even permanently distorted by comparison with the now ubiquitous Western Calendar" (1993, 339). Based on the comparison of ethnographic data gathered during the early twentieth century with data gathered in the late 1970s, Burman states about the time-reckoning practices on Simbo in the Solomon Islands: "The new time reckoning system [the Gregorian calendar] is based on an abstract, qualitatively undifferentiated, quantitatively segmented continuum, which has an autonomous existence, independent of seasonal change and the rhythm of human activities." She goes on to point out that a significant difference between the Gregorian calendar and traditional Simbo time reckoning is that the new calendar "is not the monopoly of a privileged few" (1981, 262).

As with clocks, a feature of this is the separation of the timing of a celebration from the environmental cycles that originally produced it, and this separation now exists outside of the consciousness of most people.

The calendar, instead of mediating between users and environmental cycles, now merely mediates between users and designers, even controllers, of the calendar. In this mediation, the power to control communities of practice through the objects of time they use overshadows the use of the objects of time to understand local timescapes. A good example is the holiday of Thanksgiving in the United States. It is still anachronistically conceived as a harvest festival even though it is held long after the harvest period for most temperate-zone crops. Originally, it was held in late June, but the government moved it to November. Even after the move to November, Thanksgiving was determined by the will of state governments and not by the Federal government, which resulted in its falling on different days in neighboring states. It was only in 1941 that the Federal government set the date for the entire United States. Jones used his experience of this as a source of insight into the debates about the timing of Easter:

> As I write these words in New England, Thanksgiving approaches. The President of the United States has set November 23, the Governor of the State of Connecticut November 30. In New England the dogma attached to the observance of Thanksgiving is made sensible by a ritual preserved by tradition from the elders. There is a certain spiritual and mystical element indoctrinated by types and figures, subject to different interpretations in different places. I am a pilgrim from New York State, raised in another tradition. Friends from home may undertake a pilgrimage to Connecticut to share a turkey with me on their Thanksgiving, November 23. But in these waste places in this barbarous land, children must attend school all day and miss the great mid-day feast which our tradition prescribes. A week later, when our friends from home have gone, children will be observing the holiday and keeping me from my work." (Jones 1943, 81)

So when one thinks about the assumptions embedded in calendars, one sees that calendars become cognitive tools for social coordination, and the sort of coordination that is sought determines the consciousness of time for the communities of practice that rely on the calendars. The Gregorian calendar raises consciousness of the power of government to determine holidays; the Hindu calendar raises consciousness of the effect of astronomical alignments on one's future prospects; and the Jewish calendar raises consciousness of common traditions and heritage tied to a specific place—Israel.

Objects of time determine senses of time, because they naturalize assumptions. They hide their arbitrary culturally constructed components as well as the selection of environmental cycles. But by eliminating different temporalities and their cycles, homogeneous time undermines the experience of the uncanny generated by the intersection of different cycles, calendars, or environmental rhythms. The uniform Gregorian calendar, then, through its dominance over cognition about time has, in fact, and ironically, served a secular purpose by discouraging consciousness of the intersection of different cycles in favor of the authority of a single representation of time through mass-produced artifacts. The coincidence between Christian celebrations and those of other religions that used to evoke a sense of mystery have been turned on their heads and used to suggest that Christian holidays are merely derived from pagan traditions. Whereas there might be some truth to this claim with regard to particular practices (such as the Christmas tree or the Easter bunny), the ethnoastronomical knowledge known at the time of early Christians by Romans, Greeks, Jews, and Egyptians, and the consciousness of astronomical cycles and alignments in early Christian writings, such as those of Anatolius, indicates that the Christian holidays did not replace those of other religions as much as they created Christian significance for the celestial events noted by Christians and non-Christians alike. In fact, the very narrative of the pagan origins of Christian holidays hides the astronomical significance of the timing of these holidays to both Christians and non-Christians. Like the clock and the calendar, this tale elides observable environmental cycles in favor of privileging a cognitive artifact—the calendar—which then determines how one thinks about time. A paradoxical result is that the supposedly sky-oriented pagans have had their holidays fixed to dates in the Gregorian calendar.

Expert versus Lay Calendars

A written calendar, then, is not so much as a cognitive tool to assist the reckoning of time, but a cognitive and cultural tool that can either promote social coordination or intersubjective senses of uncanniness, or even both, as in the case of the Jewish calendar. Calendars as artifacts are tools of power and social coordination. There also is an important contrast between complex calendars that require trained experts to interpret them versus simple calendars that almost anyone can use. The former are associated with astrologers. In Hinduism, the ability to chart the multiple cycles in the sky is important and involves trained specialists. In contrast, the Gregorian calendar requires very little specialized knowledge to use.

This difference between calendars does not indicate different social orders, but instead, suggests different ways in which power is instantiated by means of calendrical systems. In the case of the astrologer, the power of using time to organize significant events is in the hands of a particular religious specialist. In the case of the Gregorian calendar, the power of using time to organize significant events is embedded in the artifact itself. This allows the power structures inherent in the calendar to be hidden from view.

What uniform calendrics accomplish is the control over celebrations. Here, the homogeneous empty time in the Gregorian calendar allows states to commemorate their holidays and histories in ways that need not be tied to astronomical events. This is a break from Hindu, Muslim, Jewish, and Christian traditions. One could argue, then, that homogeneous empty time is a tool for the secular state to attempt to place the nation above religion and its rhythms—but a tool that has been largely unsuccessful. After all, in Trinidad and Tobago, half of the state-recognized holidays are moveable religious celebrations—Hosay, Eid, Carnival, Ash Wednesday, Easter, Phagwa, and Diwali. The fixed secular holidays are New Year's Day, Indian Arrival Day, Emancipation Day, Independence Day, Republic Day, Christmas, and Boxing Day.

The illusion of holidays fixed to a date in one calendar moving when represented in another suggests the awkward juxtaposition of different logics, and that this juxtaposition is obscured by how a particular calendar mediates the different logics. The manner in which the Gregorian calendar mediates the different logics of Hinduism, Islam, and Judaism is to represent holidays in those faiths as moving—a cognitively useful, even if not entirely fair or accurate, representation. Objects of time think for their users, but the manner in which the Gregorian calendar thinks for non-Christians is not always useful to those non-Christians. What is useful in one respect is not functional in another.

Chapter 4

Polyrhythmic Temporalities (Confounding the Artifacts)

The gods confound the man who first found out
How to distinguish hours! Confound him, too,
Who in this place set up a sun-dial
To cut and hack my days so wretchedly
Into small portions! When I was a boy,
My belly was my only sun-dial, one more sure,
Truer, and more exact than any of them.
This dial told me when 'twas proper time
To go to dinner, and when I had aught to eat;
But nowadays, why even when I have,
I can't fall to unless the sun gives me leave.
The town's so full of these confounded dials
The greatest part of the inhabitants,
Shrunk up with hunger, crawl along the streets.

—*The parasite's complaint from Aulus Gellius (1927, 247 [second century, 3.3.6–14])*

By now it should be apparent that the clock and the Gregorian calendar are culturally conceived tools that direct thought about time. With the expansion of European influence, they have become global. This chapter examines their coexistence with other forms of conceptualizing time to demonstrate that temporality cannot be discussed solely in terms of the clock or the Gregorian calendar even in those societies that seemingly have wholeheartedly adopted these cognitive tools. Instead, one must grapple with the intersection of multiple social rhythms and cultural temporalities. Even though the tools for reckoning time are clocks and calendars, the time they indicate does not uniformly dictate rhythms and cycles throughout the world. This produces entanglements of the logics embedded in globally distributed objects with local practices. Often, these entanglements generate strategies for reconciling social rhythms defined by objects of time and social rhythms that are out of sync with these objects.

Others have made the point that time consciousness existed before the widespread use of clocks and the Industrial Revolution (Glennie and Thrift 2002, 2005, 2009; S. Stern 2003; T. C. Smith 1986). Moreover, both the ethnographic and the historical records show that long after the widespread adoption of the clock, other modes of time consciousness remained (Birth 1999; Bruegel 1995; Glennie and Thrift 2009; Pickering 2004). Similar points can be made about the Gregorian calendar, namely, that awareness of other cycles coexists with it, without the calendar mediating the knowledge (Birth forthcoming). Thus, clocks and the Gregorian calendar coexist with other temporalities, even though clocks and the Gregorian calendar emphasize a self-referential uniformity of time.

Bachelard's concept of rhythmanalysis provides a framework for conceptualizing multiple temporalities. In the last chapter of *The Dialectic of Duration*, Bachelard (2000 [1950]) suggests that the study of time must avoid the confusion of time with uniformity, but instead emphasize the phenomenology of rhythms. Broaching the topic of the nature of time from a phenomenological perspective, Bachelard argues that rhythms are a fundamental element of all existence, and that one of the challenges facing people is the development of an awareness of the interaction of all the rhythms in which their lives are enmeshed. Moreover, he argues that happiness is achieved through living in consonance with these rhythms (2000 [1950], 152). The problem is that the timescapes of many people (and of Gellus's parasite quoted at the beginning of the chapter!) are dominated by what Bachelard calls "superimposed time"—concepts of time that are imposed on the flow of existence. To Bachelard, this creates tension between the rhythms that exist and the superimposition of artificial times that are uniform and homogeneous.

Lefebvre adopted Bachelard's concept of rhythmanalysis to analyze the "bundle" of rhythms associated with physical, biological, psychological, social, and cultural processes (2004, 9). Rhythmanalysis for Lefebvre does not separate the rhythms in this bundle to study them in isolation, but instead he identifies them and studies them in relation to one another (2004, 10). Lefebvre's rhythmanalysis is a methodological approach to study communities of practice in relationship to timescapes—to recognize and relate multiple cycles within an environment. This emphasis on relationships between rhythms forms the basis of his vocabulary of polyrhythmia, eurhythmia, and arrhythmia. Polyrhythmia is the existence of multiple rhythms; eurhythmia is the consonance of these rhythms; arrhythmia is the conflict of these rhythms (2004, 16).

Polyrhythmia and arrhythmia are common human conditions—much more so than the dominant narrative of clocks and calendars attests. In this narrative, after short periods of adjustment to the new technologies, polyrhythmia and arrhythmia disappear in the face of the ideological power of the artifacts. In fact, because these globally distributed objects of time seem to superimpose a uniform time on local communities of practice with different cultural traditions, the artifacts generate polyrhythmias and arrhythmias, and such situations probably extend as far into the past as does the use of artifacts to think about time.

The tension between objects of time and local practice is ancient. It is a consequence of the distribution of logics mediated by objects. With regard to classical Greece, Reiche states, "The coexistence of different ancient calendars reflects the coexistence of different occupations, each with distinct (though, in practice, of course, overlapping) calendric needs. Long-range fiscal planning presupposed the computation of accruals and due-dates; Hellenistic astronomy, that of eclipses and planetary motions" (Reiche 1989, 37). In Trinidad, different work patterns have different rhythms and cycles (Birth 1999). In Indonesia, Kodi time involves a "plurality of temporal scales" (Hoskins 1997, 78) and employs "notions of duration and sequence, both of which have correlates in the natural world (the movement of celestial bodies, the changing of the seasons, the waxing and waning of the Moon) and in human biological experience (the processes of aging, death, and reproduction)" (Hoskins 1997, 79). Some calendrical systems, such as those found in Hinduism, emphasize the intersection of cycles and rhythms rather than attempt to establish uniform time. Others, such as the Islamic calendar, recognize that the religious cycle will not be synchronized with seasonal changes. The Romans kept two kinds of time. Parapegmata were objects that charted astronomical and meteorological cycles (see Lehoux 2007); fasti were calendars that charted civic and ritual cycles (see Rüpke 2011 [1995]). Before Julius Caesar's reform of the calendar, these two cycles did not coincide (Feeney 2007, 198–199, 202–203), so any cycles tied to weather, such as agricultural cycles, could not be coordinated with fasti.

Looking through the ethnographic and historical record, societies that rely on small numbers of devices that can be applied anytime and anywhere to determine time or measure time are very recent and are associated with European influence and imperialism. To make uniform timekeeping the standard by which other cultural systems of time are understood is to make the recent, unusual cultural model of modernity the guide for understanding patterns in and between other cultures.

Attention to contemporary cases of multiple temporalities, such as those found in Trinidad, is useful for considering how temporalities not structured by the clock or calendar might work, and how they would be reckoned, and also for recognizing the limits of homogeneous uniform time in modernity. In chapter two, I gave a couple examples of this, but then turned to the literature on medieval Europe to further explore how multiple temporalities functioned before the clock.

Admittedly, this softens the points about the extent to which cognitive artifacts determine thought that were raised in the chapter one, but only somewhat. It demonstrates that those who are not bound to apply clocks and calendars to all issues of time can approach temporal problems in ways not mediated by these cognitive artifacts. At the same time, it sustains the point that the moment one adopts these artifacts to meet cognitive challenges, one's thought is constrained, and one's ability to address issues emerging out of polyrhythmia severely curtailed.

Even though Trinidad is a postcolonial setting, and the imposition of clocks and the Gregorian calendar there accompanied European colonialism, it is a mistake to see polyrhythmias and arrhythmias as products of colonialism. Instead, they are part of the European tradition in which efforts to standardize modes of reckoning time produced polyrhythmias. Recalling the discussion of medieval York Minster, the workday was structured by the intersection of different cycles of activity and environmental cues: the liturgical cycle of churches and monasteries, the cycle of daylight, the eating habits of vicars and workers, and the estimate of travel times to walk a "space." Polyrhythmias are not unique to Europe. They are a common feature of agricultural practice since many crops have different growing cycles. Polyrhythmias are also an issue for centralized statecraft wherever it is found: calendrically generated polyrhythmias shape politics for the classical Maya (Puletston 1979; Rice 2004); astronomers associated with calendrical calculations were government officials in Chinese dynasties (Needham 1959, 186–209; Loewe 1999), as well as in the already-mentioned societies of the Romans and Greeks.

That said, European colonialism was forged at a time of European arrhythmias. The imposition of the Gregorian calendar created confusion in Spain's New World colonies (Kelsey 1983). Since Protestant nations did not adopt the Gregorian calendar quickly, there was sustained confusion in the New World as different calendars continued to be employed. Mark Smith notes, "highly literate and informed Gregorian-style European travelers to the Julian-calendar American colonies sometimes had trouble remembering new-style dates, even though they had ready access to calendars and almanacs" (M. Smith 2001, 531). Since much colonial commerce was based on bills of exchange and credit, multiple calendars fostered financial confusion.

Even before the Gregorian reforms, there was confusion—in England the "old-style" calendar began the year on March 25, whereas the "new-style" began it on January 1. Ponko reports a case before the Privy Council in Great Britain concerning a debtor—the debt was due on January 15, 1578, but there had been no determination as to whether this date was based on the new-style calendar, or the old-style calendar in which the year started on March 25. The debtor argued that his debt was due according to the old-style calendar in which January 15, 1578, was a full year after it would have been in the new-style calendar; the creditor argued for the new-style calendar. The Privy Council decided in favor of the creditor and ordered the debtor to prison until the payment was made (Ponko 1968, 53–54).

The colonial Caribbean was forged out of capitalist polyrhythmias and arrhythmias. The cycle of sugarcane production (from one to two years, with a March–May harvest) did not mesh with the agricultural and fishing cycles of North Americans who provided food for the plantations (both cycles of 12 months with production peaking in the summer), or with systems of credit (locally variable). Mintz (1985) has argued that the time consciousness associated with industrial capitalism emerged on sugar plantations in the region, and the Caribbean has played an important supporting role in the emergence of the modern uniform time consciousness. The reliance on bills of exchange for commercial transactions between Caribbean colonies, North America, and Europe was one catalyst for calendric reform—a catalyst so powerful that by 1752 the virulently anti-Catholic English adopted Pope Gregory's calendar with little protest (Poole 1995, 1998; Alkon 1982), even though it resulted in their "losing" 11 days in September. In the British North American colonies, there seemed to be even less concern over the change and even greater support for it than in Great Britain (M. Smith 1998). In the Caribbean, the change does not even seem to have been noted—quite possibly because people in the Caribbean were quite used to dealing across the calendric divides that separated the Protestant from the Catholic European states after 1582 when the Gregorian calendar was adopted by the Roman Catholic Church. In fact, it seems that the transition to the Gregorian calendar in the Spanish colonies in 1582 caused greater confusion than the adoption of this calendar by the British colonies in 1752 (compare Kelsey 1983 with M. Smith 1998). The reason for this is that the 1582 reforms created a plurality of calendars in the New World that the 1752 reforms alleviated. Before 1752, those engaged in trade found themselves using both the Julian and the Gregorian calendars—"Colonial merchants had to know which countries, even which individual ports, used which calendar" (M. Smith 1998, 578). The adoption of the Gregorian calendar by the British resolved much of this problem and was greeted with "relief" (M. Smith 1998, 580). Europe's New World colonies, then, emerged in the context of temporal plurality.

The colonies that relied on slave labor in the New World were also timescapes quite different from those in Europe. In Europe, churches dominated the local timescapes, and even after the innovation of factories with their own time signals, there were conflicts between the workplace and the parish church (Corbin 1998). On plantations in the New World, it was the plantation itself that generated the time signals (M. Smith 1997). In the agrarian soundscape, church bells are absent but the plantations' signals are prevalent, with bells, horns, and conch shells to signal the time that organized the workday:

> Again, o'er hill and dale, resounds the shell!—
> Sharp summons to return afield again;
> Groups follow groups, and mark the signal well:—(Hosack 1986 [1879], 118)

In the Caribbean, plantation labor was structured by the plantations' self-referential time reckoning, not by the sound of public bells.

This self-referentiality was imposed on a labor population with its own diverse ideas of keeping time. The calendars and time reckoning of the enslaved African population have received little attention, although there are clear cases where they maintained calendrical traditions (see Smith 2001). After emancipation in Trinidad, the labor shortages eventually prompted the government to bring indentured laborers from South Asia—and these laborers brought Hindu and Muslim calendars. These calendars persisted, as is attested by official government recognition of Hindu and Muslim holidays.

As a result of the multiple European calendars used at the time of the emergence of the Caribbean plantation society, plantations became their own arbiters of time with a workforce that brought very different ideas of time reckoning—West African, Catholic (Gregorian calendar), Protestant (Julian calendar in the English colonies until 1752), Hindu, and Muslim. The Caribbean can be described not only as a region in which the temporal discipline associated with industrial capitalism emerged, but also as a region in which this temporal discipline was forged in a context of polyrhythmia.

The Caribbean has also produced thinkers and writers who explicitly confront multiple social and environmental rhythms.

Wilson Harris and Édouard Glissant create a Caribbean phenomenology of time that requires the experience of multiple temporalities—environmental and social, local and global. Harris called for a need to "deepen our perception of the fauna and flora of a landscape of time" (1999, 182). Glissant wrote, "Our quest for the dimension of time will therefore be neither harmonious nor linear. Its advance will be marked by a polyphony of dramatic shocks, at the level of the conscious as well as the unconscious, between incongruous phenomena or 'episodes' so disparate that no link can be discerned" (1989, 106). The Cuban literary critic Benítez-Rojo has suggested that the Caribbean is more of a rhythmical area than a cultural or a societal one. For him, rhythm is a window into what he calls the Caribbean of the senses, sentiment, and presentiment (1996, 10), where time unfolds irregularly (1996, 11) and polyrhythmically (18). Polyrhythms remain part of the social and cognitive fabric of Caribbean life and thought.

Contingent Timing

Since many individuals in rural Trinidad work in several occupations, what Comitas (1973) calls "occupational multiplicity," and since each occupation has its own rhythm, the problems faced in coordinating activities and relationships are substantial (Birth 1999). Many of these rhythms involve contingent timing. From a temporal perspective grounded in clock time—a perspective that subtly distorted my analysis in *Any Time is Trinidad Time* (Birth 1999)—occupational multiplicity fosters an image of temporal chaos, but life in rural Trinidad is not chaotic. Instead, many tasks have their own task-specific temporal signals. The cycle of route taxis coming and going allows one to time when to seek transportation, and their sequence allows riders to anticipate when their favorite drivers will be picking up passengers; the toughness of bull grass during the heat of the day determines the optimum time to cut it; the sound of schoolchildren walking by the road signals the opportune moment to open shops; and the fact that the tropical downpours can be heard before they arrive provides a signal for when to seek shelter or to cover drying cocoa.

In *Any Time is Trinidad Time*, I discussed Ranjit, a man who ran a small shop, who made and distributed candies with his wife, and who also cultivated cocoa. As I wrote, "Ranjit's life involves the intersection of multiple cycles: the commercial cycle of delivery of the candy he and his wife make, the agricultural cycle of harvesting and processing cocoa, and the daily cycles of business at his shop. Ranjit's household is like most others in Anamat, because of the intersecting cycles of agriculture, other forms of labor, and commerce" (1999, 56).

One can document that Ranjit balances these cycles, but how he balances them is more difficult to determine. The clock is of some use in how he meets his challenges. Some of his weekly activities are determined by clock time, such as the start of services at the church in which he is involved. But clock time is not sufficient even for things that seem determined by clock time, since other factors can intervene to decouple activities from the clock. The start and end of the school day is determined by clock time, and that has a ripple effect of determining when students walk to school. Since Ranjit's shop offers snacks and drinks that many older children buy, this daily rhythm of school is important for business. He must anticipate when the children will be walking by, and sometimes, particularly when school closes early during the dry season due to a lack of water, the clock is not of much use in this regard. For the snacks that Ranjit himself produces and sells to other stores, he must anticipate their sales and demand, and their sales and demand themselves vary, often in relationship to the passing of schoolchildren by their stores.

Consequently, Ranjit needs to be a student of rhythms that are not able to be represented solely by clocks and calendars. He must deal with two cognitive challenges: understanding contingent timing and developing anticipatory signals.

Contingent timing refers to tasks for which their ideal time is dependent upon some other activity cycle in the timescape. Ranjit could have his shop open at all times, but then he would not be able to drive to other shops to sell his own product or to wholesalers to restock his own store's shelves. Instead, he chooses to be aware of when he is likely to get the most sales in his shop, and to do his distribution work at other times. This is the case for all the local shops. As another shopkeeper said:

> When somebody might come in the shop here calling, "You supposed to be in the shop, you supposed to be in the shop," that time I in the shop there and have to wait hours before somebody come. I can't wait in the shop and waste my time. Right? I around the place, I find something to do.

In the case of this shopkeeper, the "something to do" included driving water trucks and a taxi.

For shopkeepers, the times they are likely to get the most sales is contingent on the behavior of others—when schoolchildren go to and from school, or when the workers return home from work. Consequently, the timing of their stores' hours is contingent upon the actions of others, and the timescape of storekeepers is as closely tied to others' activities as it is to the clock or the calendar. Indeed, as tools that mediate cognition about time, clocks and calendars often have little to offer to solve the timing challenges of Trinidadian shopkeepers and are even maladaptive during those periods when the timing of school letting out is irregular because the water supply is irregular. Devices that tie duration to timing through temporal uniformity are not of cognitive assistance in these situations, but instead, timing is tied strictly to succession and to recognizing the signs of how events unfold.

Contingent timing is most acutely felt in agriculture. Planting is an activity with contingent timing related to the weather; harvesting must balance the contingent timing of when crops ripen and the contingent timing of market cycles. The cultivation of cocoa involves its own particular challenges with regard to the timing of activities. Cocoa trees produce pods throughout the year, but the harvests typically occur at the end of the two rainy seasons. Once harvested, the beans need to be fermented and then dried. Fermentation involves covering the beans to trap moisture and putting the covered beans in the sun. In this process, the pulp liquefies and drains away, and the chocolate flavor is enhanced. Once fermented, beans are exposed to the sun and dried. Due to when the beans are harvested, fermentation and drying must be protected from the time when brief downpours of rain are still likely. To help deal with this, cocoa houses have sections of their roof on rollers to quickly move over the beans if the threat of rain arises, but this requires somebody to be in proximity to the cocoa house to move the roof when necessary. Proximity alone is not sufficient, however, since most cocoa farmers also engage in other occupations, meaning that they might be in the middle of something else when they hear the rain approaching.

In the history of cocoa production in Anamat, contingent timing was an important component of the day labor that plantations relied upon. The plantations supplemented their workforce with local farmers who had small plots of land on which they grew cocoa, and they would often take care of their own trees before seeking wage labor on plantations. C. Y. Shepherd documents this when he writes: "On most estates, the labour population is partly resident and partly casual, and is employed daily. The casual labourers usually live in neighbouring villages, or on scattered holdings, and the majority divide their energies between working on the estate and on their own land, and are subject to no control by the estate." He adds, "There is no obligation for labourers to work regularly throughout the week, nor for estates to provide regular employment, though an attempt is made to do so for resident labourers. Most of the labourers combine some other occupation, such as rice growing, canefarming, cacao cultivation, or vegetable gardening, with labouring on the estate" (1935, 86).

This led to a variety of strategies for structuring work. One elderly Indian remembered that when labor demand outside of the plantations was high, the plantations tended to emphasize wages paid by the task rather than by the day, and assigned tasks that could be completed quickly. Patron-client relationships were also important, with particular families of small farmers consistently working for the same plantation—one man remembered, "I like[d] working for M because whenever I needed work, M would find work for me. Since I work my family's land, I not always work year-round for others. Having the arrangement with M made things easier."

The profitable selling of agricultural produce in markets is aided by good timing. This was most acutely apparent to me after the attempted coup of 1990 when those who had breadfruit trees took full advantage of the high food prices in Port of Spain. Breadfruit is normally not a profitable crop, but after the attempted coup there were shortages of flour and rice. It was in August, so root crops were out of season for most people, but it was the height of breadfruit season. As a result, there were pickup trucks loaded with breadfruit leaving Anamat every morning headed for urban markets.

Anticipatory Signals

The ability to take advantage of contingent timing versus being hurt by it involves anticipatory signals—looking for clues to guide the timing of one's actions. Temporal cognitive tools, such as clocks and the Gregorian calendar, have designs that emphasize uniformity and are of limited use in such situations. The diversity of rhythms generates unpredictability only if viewed in terms of uniform durations being linked to the reckoning of time, but if one separates these cognitive tasks and recognizes polyrhythms, then the unfolding of activity is predictable. The challenge ceases to be knowing what time something will happen, and shifts to a recognition of the contrast between events that take place at a particular time of day, events that take place after specific other events, and events that take place after a particular duration. I have already noted some cases of anticipatory signals, such as the Hopi anticipation of the winter solstice discussed in the previous chapter.

One of the most powerful anticipatory cues in rural Trinidad is the sound of car engines. Early in my field research, I was amazed that almost everyone in the community in which I worked could identify engine noise with a specific car and driver. The sound of a vehicle coming preceded its coming into view, and the ability to identify cars and trucks by the sound of their motors was an important anticipatory skill. In the mornings, it gave cues as to when cars were leaving the community, and the sounds of specific route taxis going to the nearest market town not only indicated when they were leaving, but with an estimate of how long it would take them to get to the market town and return, when they would be coming back. This knowledge was an important component for schoolchildren who attended school in the nearest market town knowing when to signal for their favorite taxi driver. It was also essential knowledge for households without cars that wished to purchase propane gas tanks or cases of soft drinks, or to sell bottles to the vendors who passed through the village on a regular basis—speaking as someone who rented a house with a propane-fueled stove, I, too, came to recognize the distinctive sound of each vendor's truck. For shopkeepers like Ranjit, engine sounds were extremely useful on those days when school ended early. The sound of teachers' cars would significantly precede the arrival of children seeking to buy sweets and drinks on their walk home from school.

This elaborate choreography of activity to engine noise attunes one to the intersection of soundscape and timescape. It is not that time is solely reckoned by clocks or even by the position of the Sun. Indeed, the sound of cars and trucks created a rhythm of activity not tied to the clock at all, but a regular rhythm nonetheless. Likewise, the sound of bells tied to canonical hours and liturgical practices in medieval European towns must have also merged soundscape and timescape without residents having to directly reckon time (Glennie and Thrift 2005, 75). This suggests one other influence of the clock and the calendar: individuals consult these tools by which time is reckoned rather than rely upon public symbols. The importance of individually possessed cognitive artifacts that encode a shared model of time has supplanted collective perception of public signals (Birth 2011b). Indeed, the bells that chime on many university campuses are anachronisms, or what Tylor would call "survivals" (1958 [1871], 16), remnants of past practices that are no longer relevant. The City University of New York in which I work still occasionally refers to the scheduling matrix of classes as the "bell schedule" even though few classes begin when the library's bells chime. Behaviorally, this allows individuals to check the time whenever they want, rather than have to wait for a time signal.

Anticipatory signals are an important means for taking advantage of contingent timing. One memorable example for me came a few days after the 1990 attempted coup d'état in Trinidad, when the curfew was still in place. Police had orders to "shoot on sight" anyone on the streets after six in the evening, but since Anamat was far removed from the fighting in Port of Spain, the residents found the curfew onerous and pointless. The nearest police station was ten miles away, so the presence of police was always minimal—even more so in the wake of the coup. Besides, everyone knew the distinctive putt-putt-putt sound of police jeeps. The road followed the twisting path of the river through the cocoa walks and jungle, so as long as one was around a bend, one could easily hear the sound of the police coming from at least a quarter of a mile away—giving one plenty of time to seek shelter in the forest, or under the platform of an upstairs house—a space that was defined as "inside," as opposed to on the street. This meant that the beginning of the curfew was not straightforward. While the official start was six in the evening, many were not inside at that time, but instead took a chance to be on the road after six. These people trusted their ability to hear a police vehicle long before the police would come into view, much less see the person. In effect, the de facto start of curfew was the sound of the police coming.

There was a problem, however. The moments immediately following six in the evening were full of doubt. This was a time when some people raced home in cars in order to avoid being caught out after the curfew. But there were rumors that the police commandeered private cars—they knew the sound their jeeps made and their best sonic camouflage was to arrive in a Nissan or Toyota sedan, rather than in a police jeep. In Anamat, nobody had, as yet, witnessed such a trick.

One night, despite the curfew, a group of men gathered outside one of the rum shops in the village. They wanted to hear the stories of the coup from Reginald, a man who had grown up in the village, but had become a high-ranking officer in the protective services in Port of Spain, the capital city. He witnessed the events of the coup "live and direct." Since it was after curfew, as a precaution, the owner of the rum shop had his front closed, and was selling beer to us through a side door. In the cool night air, we stood in front and talked about the coup.

At around six thirty we heard a car engine. One of the taxi drivers with us, a man who could identify the sound of just about every engine owned by anybody within five miles, furrowed his brow. "Whose engine that?"

Another man said, "That sounds like Boog from up Orange Valley Road."

"What he doin' here after curfew?" the taxi driver said.

Reginald looked serious and said, "We better move close to the house. Police taking private cars."

Reginald, being a police officer, was a voice of authority, and we complied. Some young men liming across the road did not hear him, and continued to sit on the rusting shell of a refrigerator next to the road.

A minute later, a huge Nissan sedan pulled up. It was Boog's car, but when the doors opened, several constables carrying guns poured out of the car. They looked at the crowd of a half dozen men sitting in the carport in front of the rum shop. They seemed to note the presence of Reginald and to recognize him as a man who outranked most local constables. Acting as if they did not see us, they turned their attention to the men across the road. Through shouts and waves of their weapons, the officers ordered the men to the ground, threatened bodily harm, gave a warning not to be caught outside during curfew again, and then climbed back into the car and drove away.

Sound turns corners and passes through walls. It is an index of human activity and presence, even when the activity cannot be seen. This makes it an excellent anticipatory signal. In medieval European towns, the sound of bells could travel where there was no line of sight; in factories during and after the Industrial Revolution, the sound of the whistle could indicate the end of the workday without workers having to see a time clock. In social settings of public, collective time, the sonic signal is possibly more useful in coordinating activity than is the practice of individuals having their own timepieces. In effect, sound is an important component of many timescapes, whether it be bells in medieval towns, whistles in factories, or conch shells and horns on colonial estates. Indeed, I remember the soundscape of where my grandfather lived in rural Pennsylvania. Noon was punctuated by the siren at the volunteer fire company, but the rest of the day involved the horn of the train that ran past the small community—it would blow its horn before reaching the railroad crossing, and if one knew the train schedule, one knew what time it was based on the train's signal.

Sound is polyrhythmic and difficult to be synchronized, however. The speed of sound is slow enough for there to be a perceptible delay between the sight of a bell ringing and the hearing of the sound. If one positions oneself in a soundscape with multiple bells that chime the hour, but in a place not equidistant from the bells, then even though the bells ring at the same time, their sound reaches one's ears at different moments. An important shift in Western European temporalities has been from sound signals to visual cues to watches carried by individuals. Bells and then time balls, such as the one dropped to celebrate the beginning of the New Year in Times Square, became anachronistic as the individually possessed timepiece achieved dominance.

A Day in the Life

Many classic ethnographies that treat the reckoning of time in societies not structured by the clock contain discussions of the cues used to determine the time of day. These discussions are often sequential distillations of polyrhythms—Evans-Pritchard's classic discussion of the Nuer time reckoning refers to both pastoral activities and the position of the Sun, although his discussion of the cattle-clock is solely a sequence of pastoral tasks (1940, 101–102). While clock time is the common referent in Trinidad, activities are often not tied directly to clock time, but to other events, which leads to the following abstraction from a typical weekday during the school year. This is a context-embedded timescape in which the major indicators of time are, in fact, points where different social rhythms intersect. My abstraction of the day is from the position of a shopkeeper, simply because for short fieldwork trips I have stayed in a shopkeeper's home.

First, the alarm clock sounds. This is promptly turned off, and everyone goes back to sleep.

A bit later, one hears the first sound of trucks traveling down the road. They are farmers taking the produce to the Port of Spain market, and the noise of their vehicles in fact signals the time for the shopkeeper to wake up in order to be ready to serve customers.

Soon after the Sun rises, there is increased traffic. This is produced by people traveling to work and students traveling to school. It is important for a shop to be open at this time for customers to pick up small items to take with them.

After this comes the sounds of the primary schoolchildren walking to school, and soon after this people who "make garden" returning from their plots of land. The tardy children are still on the road when those who make garden return home, and this typically leads to these farmers chiding the children for being late to school. This is also a time when these farmers often stop off at a shop to buy a cigarette or two.

The morning hours then become quiet until noon, when one hears "Nadia's Theme"—the beginning of *The Young and the Restless*. Since most homes are open and well ventilated, and many people watch this television show, the sound of its theme music is a very effective marker of the time throughout much of the community.

At some point after *The Young and the Restless*, one once again hears the sounds of schoolchildren on the road. As is the case with the morning rush, this is an important time for stores to be open, but as mentioned, it is not always predictable—particularly in the dry season when the students are often dismissed early.

At about six PM, the parrots fly overhead.

Then, around seven PM, one hears the theme for the evening news. Like the theme for *The Young and the Restless*, this temporal cue can be heard throughout the community.

These cues are sufficient to structure most days without attention to the clock. Since houses are open to allow for ventilation, the sound of television shows drifts into the street, and the daily round of television programs and their theme songs can provide some grounding in clock time, although one should not assume that all shows start on the hour—many did not during my field research. Yet, certain shows, such as *The Young and the Restless* and *Panorama* (the evening news), did have very standard start times.

Frames of Play about Multiple Temporalities—the Case of Cricket

Cricket is a frame of play where multiple temporalities meet and the clock collides with the rhythms of play. As C. L. R. James argued, cricket is not merely about a competition between athletes, but is a competition in which national pride and national aesthetics are at stake. The West Indian style of play, which involves fast-paced, aggressive bowling and batting, is displayed in a sport that seemingly rewards patience and time management. A cricket match, whether it is a multiday test or a one-day competition, is highly structured by the clock—there is a specified start time, lunch break, tea time, and ending time. Yet, there are other factors that profoundly affect the rhythm of the game. If there is insufficient light, say due to cloudiness before dusk, then play is suspended. In addition, the game is structured around overs—an over consists of six balls bowled to one end of the pitch where the batsmen stand. Whereas there are rules about the minimum number of overs in a day for a one-day match to count, the rhythm of the overs can serve as a form of strategy, and the management of this rhythm can be used to either disrupt the batsmen or attempt to stall the game in order to achieve a draw when it looks like a win is hopeless. So overs are not measured by the clock or regulated by the clock, yet they create a rhythm within the game that then interacts with light and weather conditions.

In some circumstances, conditions of light and the rhythm of overs dictates strategy. In cricket test matches that are more than one day in length, some captains employ the strategy of the "night watchman." The role of the night watchman is not to score runs, but simply to survive the overs until play ends for the day or, if the pitch off of which the ball bounces is unpredictable due to dampness or it is breaking up, to survive until the pitch improves or is rerolled. Basically, the purpose of the strategy is to take up time and overs with a less-skilled batsman so that either the next day or when the pitch improves a more-skilled batsman can start fresh.

Whereas dynamic modeling of this strategy suggests that the use of the night watchman is ineffective (Clarke and Norman 2003), in local test cricket in the West Indies, it has great benefits due to the interactions of light, weather, and pitch. My field site in rural Trinidad is on the edge of the rainforest, and bursts of rain are common and can suspend play. This results in the classic "sticky wicket" off which the ball careens unpredictably for the batsman, and for those who are familiar with baseball and not cricket, there is no strike zone in cricket, and consequently, no "walks" awarded as a result of wild balls. Thus, a wet pitch is a liability for a batsman, and it is a common occurrence in Trinidad. Often, there are disputes about when to return to play after a shower, and umpires are usually local and thus are easily accused of local biases. Under such circumstances, a wise captain often saves the best offensive batsmen and puts in the most effective defensive batsmen. Some relish this role—one person told me proudly of one of his successes as night watchman. The wicket was damp and uneven, and the bowlers were bowling balls with great velocity and unpredictable bounces. In recounting the story, this man stood up and demonstrated what he did:

> Any ball, I go down like so [he went down on one knee with both legs directly behind his imaginary bat and with his hands cocked at an angle to drive the ball into the ground directly in front of the bat]. Next bowler, I still go down like so. I remain until dark, going down like so, with maybe a run here and there.

Conclusion

The strategies for dealing with multiple rhythms that cannot readily be represented by the clock do not consist of a single tool, like the clock or the calendar, to relate these rhythms, but instead are context-dependent cognitive strategies. Some of these strategies are based on contingencies—the sequence of events and their timing in relationship to one another is determined, but the duration of the sequence is not. Some of these strategies are based on epistemic actions—those actions described in chapter one that make problem solving more efficient, such as tapping one's foot to keep time when playing music. Other strategies are based on anticipatory signals—a known duration of an activity that is used to judge when to undertake a subsequent activity. Referring back to the discussion of medieval time reckoning in chapter two, there is a similarity not in the specific signals and tools, but in the overall context dependency of time reckoning. Polyrhythmia seems to encourage context-dependent solutions.

What, then, should be made of the temporal uniformity that has been attributed to modernity, whether it is in Thompson's industrial labor, Foucault's discipline, or Anderson's homogeneous empty time? These concepts do not capture the diversity of thoughts about time that one can find ethnographically. Instead, through the domination of context-independent cognitive tools, they seem to constrain the diversity of thinking about time in social theory.

Indeed, they seem to skew thought toward a particular epistemology that is grounded in Newtonian mechanics. The abstract, uniform categories that made Newton's physics scientific and positivist seem to still hold an attraction, even if the surface positivist epistemologies are rejected. With regard to time, it is now social theory that holds on to positivist absolutes, and physics that embraces relativity. The contingent timing and anticipatory signals so necessary for social functioning in rural Trinidad suggest that there is a need to make time contextual and processual and situational, just as other issues, such as identities, have come to be viewed in these terms. Time cannot be merely a chronological grid by which we cognitively organize events, but must instead be viewed in local and contextual terms. Time is embedded in and constituted by social life. The clock and calendar become artifacts that assist thinking about time in particular ways, not essentializing devices. This, then, is not an argument against cognitive artifacts of time, but is an argument in favor of recognizing their contexts and uses, as well as being an argument in favor of recognizing their limits.

This is not simply a matter of ideas, but it is also a matter of where the cognitive and the material worlds meet. Cognitive artifacts are material objects, and many of the cycles in timescapes are material as well. The body has a material existence, as do the bodies of other animals and plants. The materiality of the world is enmeshed with time, and our temporal devices do not intersect with this materiality in any straightforward way. Instead of mediating between mind and world, these devices mediate between mind and the minds of their dead designers and how they thought about the world.

Here, it is the forces that unquestioningly rely on these devices that turn out to be conservative, and science that is potentially revolutionary. Science's understanding of time and space has changed considerably since the emergence of the mechanical assumptions on which our timekeeping devices are designed. These new insights can be added into interpretive frameworks, rather than suppressed. Without doing so, odd suppressions of material realities emerge, such as that talked about in the next chapter, which explores how social theory has suppressed the globe in globalization.

The superimposition of clocks and calendars under modernity does little to address challenges raised by polyrhythmia. Not only do multiple temporalities persist in the face of efforts to use cognitive artifacts to superimpose specific temporal ideas, but the temporal uniformity embedded in these cognitive artifacts is ill suited to address social rhythms that are related yet not uniform.

Chapter 5

Globeness: Time and the Embodied, Biological Consequences of Globalization

The flattening of space defies Einsteinian curvature or quantum expansion but reflects the triumph of a populist and mechanical vocabulary of progress. Travel around the so-called village of the globe is made easy, swift and accommodating. Yet there lingers an unspoken apprehension of an incalculable price to be paid in pollution, in the extinction of species, and in other elemental implosive cycles which leave their shadow upon the psyche of nature.

—*Wilson Harris (1999, 62)*

The objects of time associated with modernity channel thought toward a distinctive temporality and the following features: uniform, homogeneous, and empty time; a combination of the cognitive processes of measuring duration with determining moments in time; and a mediation by the artifacts that hides the separation of temporal algorithms from environmental cycles. Whereas the previous chapter explored social polyrhythmias and arrhythmias, this chapter looks at how the logic embedded in clocks and calendars gets embodied and creates physiological arrhythmias. This is a consequence of how the objects of time encourage a conceptual reshaping of the globe and a reimagining of the body and its cycles in this age of globalization.

The reshaping of the globe involves widespread temporal coordination across multiple time zones to the time of a few politically and economically significant cities. Sassen's (1998, 2000) conclusion that there is a small set of global cities around which much of the flow of information, political power, and finance is centered leads to a realization that most of the world adapts to the timing of activities in places such as New York, London, and Tokyo. This accommodation is made possible by means of clocks and calendars. This reorientation of a multitude of local times to a few cities has effects on the body that only incompletely suppress awareness of biological cycles. Globalization and locally embedded biological cycles are in arrhythmic relationships because globalization privileges the time and cycles of only a few locations.

To make this argument, this chapter weaves together scientific knowledge about human biology and social scientific knowledge about globalization against a backdrop of the Earth and its cycles. There is an obstacle in making this argument, however: Newton's legacy of abstract time is that the science through which we understand the globe and the human body is shaped by the very cognitive artifacts used to enable global coordination and to diminish the importance of most local times. Consequently, our understanding of circadian rhythms tied to the Earth's cycles is mediated through the lens of homogeneous, uniform time rather than directly through relating circadian rhythms to the Earth itself.

Despite the Earth being a rotating globe that revolves around the Sun, ideologies of modern capitalism, reinforced by scientific and social scientific theories, treat the Earth as if it is flat. This view is affirmed by journalist Thomas Friedman's (2005) book on globalization called *The World Is Flat*, in which he argues that information technology connects people and places in ways that negate distance.

As I argued in the previous chapters, scientists and technocrats have created a global standardized time reference consisting of hours of equal duration, days of exactly 24 of those standardized hours, time zones, and a "prime" meridian that runs through the Royal Observatory, Greenwich. These units are then related to a calendar that had a religious origin, but that now has taken on a global and secular character. These logics are embedded in clocks and in the Gregorian calendar. Clock time organizes and represents labor, and with the advent of artificial light, and, later, global telecommunications, work schedules are no longer tied to daylight: both night-shift work and flexible schedules that transcend day and night are increasingly common. The Gregorian calendar homogenizes time so that seasons are of no consequence for reckoning—an important feature for a calendar that applies to both the northern and southern hemispheres. The calendar is tied to the clock because the day is now defined as 24 clock hours, not by the rotation of the Earth. But as argued in the previous chapter, the effort to achieve temporal standardization and uniformity has not done away with the polyrhythms of life, but instead has created arrhythmias.

Thus far, I have argued in favor of the recognition of multiple temporalities and a sensitivity to the collision between multiple temporal cycles, and the homogenizing character of cognitive artifacts in the form of calendars and clocks. Giving voice to the struggles against globalization's temporal homogeneity, many Caribbean thinkers express unease at treating the Earth as if it were flat, because doing so is a denial of rhythms tied to the landscape. In the quote at the head of this chapter, the Guyanese writer Wilson Harris notes the price to be paid by this flattening of space. In writing about another Caribbean thinker, Dash notes Glissant's writing expresses "unease" with "the orderliness of the European landscape, the rhythmic measure of changing seasons" (Dash 1995, 12). Glissant's Caribbean landscape is turbulent and explosive—a place of hurricanes and volcanoes, and a place removed rhythmically and temporally from Europe (Glissant 1956). Benítez-Rojo prefers to refer to the Caribbean as a "rhythmical area," rather than to define it in terms of space (1996, 75). Finally, and coincidentally, it was in Trinidad, including the village in which I worked, that Colin S. Pittendrigh, the pioneering chronobiologist, first noted and studied circadian cycles (Birth 1999, 38–39; Pittendrigh 1993, 22–23). Whereas Columbus did not prove the Earth was round by arriving in the West Indies—that was already argued by Aristotle (1936 [350 BC], 28)—West Indian thinkers and those who have studied West Indian temporalities are sensitive to the experience of local temporalities and the rhythms of the Earth as a globe rather than as a flattened space.

These people are not alone in recognizing the significance of globeness, that is, living on a rotating, revolving globe. In an era of ideologically asserting temporal uniformity to create an illusion of flatness, it seems many have recognized the temporal advantages conveyed by globeness. While Friedman asserts that the world is flat, his sources suggest otherwise. For example, the president of Johns Hopkins University told Friedman:

> I have just learned that in many small and some medium-size hospitals in the US, radiologists are outsourcing reading of CAT scans to doctors in India and Australia!!! Most of this evidently occurs at night (and maybe weekends) when the radiologists do not have sufficient staffing to provide in-hospital coverage. (2005, 16)

Night in the United States is day in India and Australia. Comparative economic advantage is normally thought of in terms of resources or the characteristics of the labor force, but globalization allows strategic use of time differences between service providers and clients to become a source of advantage.

Global and Local Times in Biological Perspective

One's place on the globe not only positions one in space, but also in multiple systems of time. These multiple times include clock and calendar time that are defined according to one's time zone and relationship to the international date line, solar time defined by the cycle of day and night, and cycles of social activity—our position in space creates a timescape of polyrhythmias. Our temporal artifacts concentrate our attention on clock time and calendars and attempt to entrain social activity to the manner in which these artifacts represent time. Human biology, sensitive as it is to both social and solar cycles, seems potentially caught between the time as indicated by our temporal artifacts, the cycle of daylight, and social activity. The clock defines some, but not all, social times, although in the coordination of global relationships, the clock structures relationships between multiple locations. In the nexus of the global and local, the potential conflicts between biology, clock, Sun, and sociality can become significant.

With regard to human biology, position on the globe is a critical variable that has been understudied due to the prevalence of laboratory work structured by calendar and clock time, which neglects seasonal and solar cycles. One's longitude indicates relationships of local solar times to time zones, which define clock time for state-specified ranges of longitude. One's latitude, while affecting local clock time only in terms of the decision to adopt daylight saving time, affects one's relationship to the temporal markers of dawn and dusk and to seasonal changes. In many animals, humans included, the further one is from the equator, the greater the seasonal variation in hormonal cycles, particularly melatonin (Bronson 2004; Foster and Roenneberg 2008; Schwartz et al. 2001; Wehr 2001). Condon (1983) ethnographically documented this in *Inuit Behavior and Seasonal Change in the Canadian Artic*, in which he found seasonal variations in physiology, social stress, activities, and birth rates. If place-bound identities are becoming more significant, as Harvey (1993, 4) argues, then one wonders about the ways in which time and space are experienced both locally—such as the tie between landscape and its rhythm that is important in the Caribbean—and in relationships that span different time zones. This experience is full of potential conflicts between global and local schedules, and between the timing of global relationships and the cycles of one's locally embedded biological rhythms.

The exploration of this topic involves relating space and time as well as sociality and biology. To explore the intersection of these issues involves theorizing about time and timing, but this task goes against inclinations in much recent scholarship in the humanities and social sciences. Discussions of postmodernity emphasize the issue of space often as a means to eschew temporality. Theories of capitalist production and exchange prefer to avoid the issue of timing in favor of the concept of "average time." Both biological and social science unreflexively adopt the standardized 24-hour day and clock time—both have developed based on the logics of time inherited from the Enlightenment (Hassan and Purser 2007, 7). Chronobiology does so without reflection about how this standard is a cultural construction; social science does so without reflection about how historically and culturally unique this standard is—as I argued in the previous chapter, the exception is used as the standard for analysis. This chapter, then, must involve a discussion of postmodernist treatments of time and globalization, chronobiological science, and the epistemological divide between natural and social science in order to develop an approach that integrates globalization, social times, and human biology. In all of these cases, there is a common thread of the persistent neglect of the consequences of living on a globe in favor of the cognitive assumptions in clocks and calendars. Once these consequences are considered, the contradictions and conflicts of time-space compression become apparent, and so does the need to appreciate these issues from a perspective that can integrate biology, geography, social systems, and culture.

The Postmodern "Elimination" of Time

Massey (1992) points out that many uses of the concept of space define it in opposition to time (1992, 68), and in so doing, eliminate the temporality of politics from conceptualizations of space (1992, 66). She cites Laclau's *New Reflections on the Revolution of Our Time* (1990) as an important example of this tendency. Laclau asserts, "[A]ny repetition that is governed by a structural law of successions is space" (1990, 41). Space forms a closed system, then. But repetition is never simple duplication (Deleuze 1994), or, as Benítez-Rojo argues, "every repetition is a practice that necessarily entails a difference" (1996, 3). Abstract space captures difference no better than uniform time does. Consequently, time, even when cyclic and repetitive, implies some form of disjuncture, if not disruption, in a specific location. Massey concludes that such disruption is necessary for the possibility of politics.

Part of the postmodern suppression of time in favor of space is related to the problem of thinking about the local in relationship to globalization. Robertson hints at this when he writes, "Interest in the theme of postmodernity has involved much attention to the supposed weaknesses of mainstream concern with 'universal time' and advancement of the claim that 'particularistic space' be given much greater attention" (1995, 26). The postmodern tendency to use universal time as a foil against which the particular can be represented is premised on the centuries-long process of representing time as homogeneous and divorced from place—clocks, calendars, and time zones have all been efforts at erasing place in the service of temporal standardization. As Adam points out, "Standard time and world time are essential material conditions for the global network of communication in both information and transport" (Adam 1995, 114). The contrast between universal, global time and locality is an unbalanced parallelism, however—there is no concept of time linked to the local. Even from postmodernist perspectives, there is a need for awareness and understanding of the interaction of universal time and local temporalities, but unfortunately this interaction has not received much attention. Postmodernist attempts to erase time, then, must be viewed as directed at artifactually mediated uniform time, but postmodernist criticisms tend to be trapped by the very time they seek to eliminate because they cannot conceptualize time outside of how it is represented by cognitive artifacts.

Castells's representation of the network society asserts "Capital's freedom from time and culture's escape from the clock" (2000, 464). In contrast to Castells's perspective, Harvey emphasizes the "annihilation" of space as the motivation and consequence of what he calls time-space compression (1989, 258, 1990, 425). Time-space compression is the consequence of the ability to send information across vast distances almost instantaneously. It creates an image of distance being able to be transcended instantaneously, thereby fostering a sense that both space and travel time are compressed into simultaneity. Adam describes time as having become "thoroughly relativized: nighttime in Wales is daytime in Hong Kong and evening in California. This knowledge forms part of the daily interactions of a global community of airline staff, financiers, business people and politicians," and she adds, "Space, it seems, is no longer an obstacle to communication; in such instances, at least, it has been rendered almost irrelevant" (1992, 177). The politics of defining time zones supports Harvey's and Adam's perspectives more than that of Castells. Hongaldarom describes a debate in Thailand over advancing their time zone to be in the same zone as Singapore, Malaysia, and Hong Kong. Hongaldarom says, "A reason for the move was that the country's time zone would then be the same as those of Hong Kong and Singapore, making financial interactions between Thailand and these two powerhouses easier" (2002, 345). In the Canadian territory of Nunavut, populated mostly by Inuit, time zones have been a particularly contested issue. In 1999, the government of Nunavut sought to have a single time zone for an area that had previously had three. There was widespread resistance to this plan: "The result was that there was no single Nunavut time zone, and even within some individual communities there was no standard time" (P. Stern 2003, 147). According to Stern, the time zone proposal desynchronized many business and hunting rhythms from governmental rhythms. The result was three changes in the time zones in an 18-month period (P. Stern 2003, 147).

One reason for the futile struggle to use time zone definitions to ameliorate national, local, and economic temporalities is the avoidance of any references to "nature." In describing postmodernism, Jameson notes that it "is what you have when the modernization process is complete and nature is gone for good" (1991, ix). This has a disconcerting resonance with Newton's rejection of observable environmental cycles as the measure of time (1934 [1687], 7–8). In rejecting these cycles, as Newton advocated, local solar time is a conceptual casualty. Have the globe and solar cycles really been eliminated, or is Jameson referring to an act of suppression? Furthermore, if human biological circadian cycles are cued by the Sun, then making solar time irrelevant also asserts that those biological rhythms are irrelevant. It is a rejection of the human species' status as a diurnal mammal that follows upon the rejection of a physical characteristic and astronomical orientation of our planet. As the previous chapters argue, it is not some new postmodernist epistemological critique that innovated the suppression of nature—this idea was critical to our modern objects of time and embedded within positivist epistemologies. For the construction of knowledge, postmodernist critiques adopt the same artifactually mediated time that is crucial to the scientific construction of knowledge.

Such rejections of nature, and consequently of time not mediated by artifacts, compromise the ability to understand the embodied experiences and material dimensions of globalization, and consequently the ability to explore specific instances of how global power relations interact with human biological processes.

Adam criticizes globalization theory from a feminist perspective and notes that the emergence of contemporary concepts of time have "lessened some of the human dependence on, but not overcome our rootedness, in the rhythmicity of the cosmos, the seasons and the times of the body" (2002, 15). While much of postmodernist theory acts as if nature, and time with it, has been eliminated, it does not feel like it has.

In the Caribbean, a region hailed as among the most postmodern of all regions (Benítez-Rojo 1996), nature's temporal influences are present and culturally asserted in the face of globalization and the homogenization and standardization of time. In my attempts to understand time and the Caribbean, I find: Caribbean writers who turn to the local rhythms of the landscape to resist this ideology; cocoa farmers who also work for the government or factories and develop strategies to minimize conflicts between these two different types of cycles; and Trinidadians who learn quickly to calculate the time differences between themselves and family who live in New York, California, or England, before making telephone calls. Because of the global distribution of family and friends, Trinidadians even learn to adjust to daylight saving time, despite Trinidad and Tobago not adopting it. There is also a sense that in global interactions, Trinidadians are aware of and adjust to the time of global cities like London and New York much more than non-Trinidadian Londoners or New Yorkers are aware of or adjust to the time in Trinidad. "Well, of course," a cynic might say, "New York and London are more influential than Trinidad," and this is exactly my point—they are so influential that they determine how objects of time are used elsewhere.

Adam argues that "clock time, world time, standard time and time zones have become naturalized as the norm vastly increases the difficulty of recognizing the role this created time plays in everyday life. Other temporal principles fade into the background. They become invisible" (2002, 17). Despite this invisibility, they remain felt. Time has not disappeared, but instead there is a conflict between a logic of temporal flexibility that seeks to make lived time and the experiences of duration, sequence, and daily cycles irrelevant. This conflict includes the contradictions between artifactually mediated times and biological circadian cycles. Lefebvre describes as the "bitter and dark struggle around time and the use of time" (2004, 74)—a struggle over how "[s]o-called natural rhythms change for multiple, technological, socio-economic reasons" (2004, 74). The language of Adam and Lefebvre suggests that time conflicts have been pushed into the dark recesses of theory and globalizing power relations, by implication making people such as Caribbean intellectuals, Trinidadian cocoa farmers, and global technology companies the inhabitants of these recesses. The suppression of embodied temporal conflicts in favor of artifactually mediated time is part of the postmodern condition.

The Embodied Experience of Time-Space Compression

The time-space compression associated with globalization creates a paradox of locations with different times vis-à-vis their local clocks being linked to the same clock time vis-à-vis exchange and the movement of information—such as the new temporal commonality Hongaldarom describes for Thailand and Hong Kong, or the fact that all of China is within a single time zone, or the policy that placed the space shuttle in the same time zone as Greenwich regardless of where the shuttle was in its orbit.

The emphasis on the instantaneous flow of information in time-space compression has not only obscured the features of living on a rotating globe, but through temporal artifacts it has also led to an emphasis on homogeneous time over the varying experiences of one's body and its immediate surroundings. The disembodiment of time-space compression from the physical and psychological temporal experience of the body affirms the denial of physiology that is crucial for the global functioning of capitalism.

An example of the interaction of time-space compression, physical discomfort, and chemical intervention was on public display at the opening of the 2004 Major League Baseball season. In April 2004, the New York Yankees and the Tampa Bay Devil Rays played the first game of the season in Tokyo, Japan. This game was an effort by Major League Baseball to showcase baseball internationally. The game also was used to promote the idea of a Baseball World Cup to parallel soccer's World Cup tournament, with some even dreaming of there one day being a global baseball league.

As Major League Baseball has global dreams, the players and fans were reporting something different. Fans in New York grumbled about the game beginning at 5:00 AM eastern standard time. The day after the Yankees returned to New York, the *Daily News* reported Yankee outfielder Gary Sheffield saying, "Man, I still don't know what time it is" (Borden 2004, 67). The newspaper story then added:

> Sheffield still felt out of it yesterday and he wasn't the only one. Tony Clark's eyes had rivers of red running through them and Derek Jeter said he still couldn't get back on a normal sleep schedule.
>
> "I might even go drink some coffee today," Clark said. "And I never do that—it messes with my stomach and everything—but I just need the caffeine."
>
> Jason Giambi said his surgically repaired knee felt fine, but the rest of him felt "like a train wreck."
>
> "I think everybody was dragging a bit," Giambi added. "I think the jet lag really hit." (Borden 2004, 67)

This time-space compression worked to make somebody feel "like a train wreck" and inspired a noncoffee drinker to drink coffee for the caffeine even though he expected the coffee to make him sick.

Speed of transportation technology would not cure this problem—the idea that it might is evidence of the conflation of the cognitive processes of judging duration and determining timing encouraged by temporal artifacts. Even if supersonic commercial flights were still available, when the Yankees played in Japan, their bodies would still manifest their New York circadian cycles and any Japanese opponents would still manifest Japanese circadian cycles—the time chosen to play the game would give one team a significant advantage. This effect has been documented with regard to games played between teams in the eastern and Pacific time zones in the United States. In the 1991–1993 seasons, home teams in the eastern time zone had a statistically significant advantage in the number of runs they scored when playing teams that were based in the Pacific time zone (Recht, Lew, and Schwartz 1995). As anyone who has experienced jet lag knows, along with luggage, one also travels with the cycles—both social and biological—of the place from which one comes. In addition, studies of jet lag have long demonstrated that it is worse if one travels eastward (Klein and Wegmann 1974).

The relationship of place and time, then, is psychologically, socially, and biologically encoded. Flows of goods, services, people, and capital across time zones should make apparent the temporal differences between places, and, in fact, many people exhibit a great deal of practical knowledge of how to deal with such temporal obstacles, yet this issue is theoretically underappreciated. This is, in part, due to the history of cultural concepts of time reckoning that have tended to bring different cycles and processes into the single, homogeneous representation of clock time.

To relate this to the themes developed in previous chapters, the artifacts of modern timekeeping homogenize times—solar time, longitudinal time, and biological rhythms. This homogenization pushes these different temporalities out of awareness in favor of the logics built into temporal artifacts. At the level of daily experience, multiple rhythms of activity are subsumed under clock and Gregorian calendar time. For instance, what I found in rural Trinidad was that the activity cycles of farming, shopkeeping, schooling, and transportation did not coincide (Birth 1999). Each of these activities had its own rhythm and pace. Clock time creates a false homogeneous representation of these different temporal cycles and activities—including cycles such as when bull grass is easiest to cut with a machete or when the parrots return to their roosts. The use of many temporalities, including astronomical cycles and the rotation of the Earth, were casualties of the need to standardize time, and the need to standardize time was a necessary condition for converting labor into a commodity.

Today, clock time is determined by the coordination of the time indicated by atomic clocks distributed across the world. The communications technology that enables time-space compression is also necessary for contemporary determinations of accurate clock time. In this global time grid, the multitude of local, longitudinally determined solar times is reduced to a relatively small number of time zones. Every meridian has its own solar time, but the clock time of a location is based on its time zone. Time zones represent a step toward time-space compression in a subtle way, in that they erase regional spatial differences—the times of all longitudes in a time zone are represented by the time at a single longitude. The determination of zone-defining longitudes is the result of national and regional political and economic processes, and time zones are easily divorced from any reckoning of solar time. The processes of ignoring the differences between solar time and longitudinal time, of averaging the length of the solar day to create mean time, of ignoring variable cycles of daylight in favor of regular clock hours, of creating time zones, and then, of making these definitions of time internationally accepted were important features of both the scientific and industrial revolutions. The further de facto consolidation of the different time zones into a much smaller set associated with globally important cities is a consequence of globalization and time-space compression.

Natural science is not immune to these processes. Indeed, as discussed in previous chapters, the concept of homogeneous, uniform time was a product of the emergence of modern science and associated with its cognitive tools. Even though theoretical physics challenges this idea of time, it is quite resilient in scientific approaches outside of physics. Homogeneous, uniform time is adopted, even demanded, as a standard of measurement in the natural sciences. In the social sciences, this concept of time has been used as a means of documenting the historically and culturally specific ideological constructions of industrial capitalism. Not surprisingly, those different views contribute to the epistemological divide between these disciplines, yet in these seemingly opposed reactions to contemporary concepts of time, there is a common rejection of the relationship of time to the globe and an emphasis on time in relationship to cognitive artifacts that represent homogeneous, uniform time. This divide creates an obstacle to both natural and social science grappling with the biological consequences of globalization.

Epistemological Divides

Elias complains that the different developments of the human and natural sciences have created difficulties in studying time: "By making the two fields appear not only as existentially different, but also as faintly antagonistic to and incompatible with each other, this type of conceptualization quite effectively closes the door to enquiries into the problem of the relationship between what we call 'nature' and 'society'" (1992, 86), and "[i]f one explores 'time' one explores people within nature, not people and nature set apart" (1992, 97).

The challenge of incorporating biological knowledge into a social analysis involves understanding the cultural assumptions on which it is built and to ascertain the effect of those assumptions on the creation of knowledge. In the case of the field of chronobiology, as time has been standardized through objects of time, it has been adopted in techniques to measure biological cycles. Yet, chronobiology also has embraced relative timing, as well. So while many experiments are organized around clock time, many also investigate the phase relationship between an environmental stimulus, such as light, and the body's response. These two different senses of time—relative to the environment versus abstract and absolute—permit a discussion of chronobiology in terms of how its thinking about time is shaped by cognitive artifacts, and in terms of how its attention to the biology and the environment provides a foundation for critiquing the role of those cognitive artifacts in the scientific construction of knowledge, and in terms of how knowledge about interaction of biology and the environment display the breadth of the consequences of the globalized artifactually mediated times.

Chronobiological studies have provided insight into both how the body's processes work together and the way in which these cycles can be disrupted. People have been placed in caves or bunkers to isolate them from time cues, with the purpose of finding their "natural" free-running biological cycles of activity, rectal temperature, and later hormonal levels (Aschoff 1965; Siffre 1975). Many species of animals have been subjected to observation, blood tests, fecal samples, urine tests, and saliva tests to evaluate how they responded to timed noises, timed lights, no time cues, movement across multiple time zones, changes in the timing of cues, and attempts to create cycles much shorter or much longer than 24 hours (see reviews of this literature in Aschoff 1981; Dunlap, Loros, and Decoursey 2004; Foster and Krietzman 2004; Moore-Ede, Sulzman, and Fuller 1982; Palmer 2002; Pittendrigh 1993; Wever 1979), and this field has grown now to include two scholarly journals devoted to it. These studies have contributed greatly to the understanding of the relationship between the environment and biological processes, particularly in the study of zeitgebers.

"Zeitgeber" is a term that refers to stimuli that synchronize biological rhythms. Cycles of light and darkness are zeitgebers. The robustness of the 24-hour circadian biological clock in humans is now well established (Czeisler et al. 1999). This is true even under unusual circumstances. For instance, astronauts on the space shuttle maintain most aspects of the 24-hour cycle, with the exception of the duration and quality of sleep (Monk et al. 1998), and sailors following an 18-hour sleep/wake cycle on a US Navy submarine maintained a 24-hour melatonin cycle despite the 18-hour cycle of activity, although this study did not examine cycles other than melatonin (Kelly et al. 1999). The robustness of biological rhythms produces conflicts between body processes and labor demands. The US Congress even commissioned a study of the effects of night-shift work on health (US Congress, Office of Technology Assessment 1991), and this literature has grown very large with Costa et al. reporting that, in their review, they found over 1,000 references on the topic (2004, 837). The effect of these work patterns on productivity and safety has spurred a new breed of consulting firm, such as Circadian Technologies Incorporated, that specialize in applying chronobiology to increase shift-work productivity.

The knowledge gained from studies of the interaction of zeitgebers with circadian rhythms during night shifts has shown dissonance and desynchronization between human biological cycles and labor demands organized by the clock and humanly created zeitgebers. In effect, globalization combined with artifactually mediated ideas of uniform time have fostered social and biological arrhythmias.

As useful as this knowledge has been to spur discussions about the effects of shift work on human health, it is still limited by a tension between methods that control for environmental variability and the study of circadian systems of organisms that cope with environmental variability due to seasonal changes in light cycles. Chronobiologists regard light as one of the most important environmental factors in regulating circadian cycles (Czeisler and Brown 1999; Czeisler and Wright 1999; Czeisler et al. 1999; Shanahan, Zeitzer, and Czeisler 1997; Wehr 2001; Wright et al. 2005). This knowledge has been gained predominately by studies of biological rhythms conducted in controlled environments in which the timing of light/dark cycles is defined in relationship to the clock. This raises a methodological problem, however: the transition from night to day is different than the transition from dark to light in the laboratory. Danilenko et al. write, "The natural zeitgeber, the dawn and dusk signal, is an obvious but little investigated paradigm given the standard laboratory procedure of rectangular wave (on/off) illumination" (2000, 438). The light intensity of dawn is sufficient to entrain circadian cycles (Danilenko et al. 2000; Meijer and Schwartz 2003). This implies that organisms synchronize to light/dark cycles, not to 24-hour days—as discussed earlier, the concept of the 24-hour day is a cultural creation with a long, complex history.

Since light/dark cycles are the most important influence on circadian rhythms, it is not the Earth's rotation, per se, but the timing of sunrise and sunset that are the relevant variables, and these do not occur at 24-hour intervals but change daily, and the time of these events varies according to season, latitude, and longitude. The longitudinal variation is further complicated by the consequences of artifactual mediation, time zones, and in some locations, the use of daylight saving time. There is considerable difference between the earliest sunrise at 5:07 AM eastern daylight saving time, and the latest sunrise at 7:14 AM eastern standard time in Boston in 2006 (the location chosen as the reference point for *The Old Farmer's Almanac*), and whereas the change to daylight saving time equals one hour on the clock from one day to the next, it is not a 1-hour change in the time of sunrise—in Boston it was a 58-minute difference in the time of sunrise.

The standardized 24-hour length of the day documented by the clock and calendar and employed in the laboratory overshadows the local and seasonal variability in the distribution of daylight across those 24 hours. Even in the approach of science to the creation of knowledge, the globe has come to be treated as if the experience of time is the same everywhere, as if the globe is, in fact, flat. This is a consequence of the homogeneous time that controls the algorithms of cognitive tools of time combined with the ability to control light cycles through artificial illumination.

Despite this, most studies of biological rhythms, even those outside the laboratory, continue to make reference to clock time—in effect, to have their consciousness of time defined by cognitive artifacts, even though the more relevant indicators of light/dark cycles are sunrise and sunset. Even the literature on transportation accidents, most of which occur in an outdoor context, tends to represent the timing of accidents in relationship to the clock, and not to the Sun (see Åkerstedt and Folkard 1995, 1996; Folkard 1997; Folkard et al. 1999), with one study that encompassed 17 countries of differing latitudes and longitudes (Adams-Guppy and Guppy 2003) ignoring how location can affect the duration and timing of daylight. If sunlight is an important zeitgeber, its neglect is only permissible with populations that live and work solely in conditions of artificial light, and who are not exposed to the entraining power of natural sunlight at dawn, but our cognitive artifacts push us to think in terms of temporal uniformity.

The emphasis on the stability of circadian biological cycles measured and charted by hours and minutes over the latitudinally and seasonally variable light/dark cycles creates a false image of a 24-hour biological clock that remains unchanged throughout the year. Presentations of biological "knowledge," then, are consequently refracted through artifactually mediated uniform hours defined in relationship to midnight, not in relationship to the variable cycles of the solar day.

Similarly, medications are given at regular, clock-determined schedules, rather than according to metabolic cycles (Zerubavel 1979), and the significant physiological processes of childbirth are mapped by the calendar and the clock, so that "[t]he more intrusive the obstetric assistance, the more the woman is forced to oscillate between the all-encompassing body time of her labour and the rational framework of her clock time environment" (Adam 1995, 49).

Social scientists often ignore or downplay biological cycles in their discussions of different cultural concepts of time and the links between temporal frameworks and power relations. The emphasis has often been to show the historically and culturally contingent nature of contemporary, clock-driven timekeeping, and paying attention to natural and biological cycles does not seem relevant. In some cases, when "natural rhythms" are invoked, they are dismissed. For instance, Castells recognizes that the network society's effort to "escape from the clock" is at odds with the close connection between the "rhythm of human life" and the "rhythms of nature" (2000, 276), and he suggests "that the network society is characterized by the breaking down of the rhythms, either biological or social, associated with the notion of a life-cycle" (2000, 277). Discussions of cultural conceptions of time, then, tend to be refracted through the representation of cultural diversity, and consequently ignore pancultural tendencies to acknowledge and organize social relationships around day, night, and biologically driven cycles of sleep and activity.

Consequently, de-globed artifactual atomic time with its standardized 24-hour day has come to be viewed as "natural" by scientists who examine human biological flexibility and variation, and de-globed concepts of time have permitted social scientists to study time as culturally constructed without having to examine the relationship of these constructions to local environmental cycles or human biology. Both scholarly communities rely on methodologies that erase nature to construct knowledge, although the social sciences admit their dismissal of nature more openly than do the physical sciences. This erasure is a de facto reductionism—multiple temporalities are elided and time becomes constituted by its artifacts. Social science is just as guilty of this reductionism as the physical sciences are—maybe more so, because physics, at least, recognizes temporal relativity, whereas the social sciences avoid that concept.

Instead of arguing against reductionism as a justification for ignoring science—a standard social scientific stance on the topic—we need to grapple with the knowledge science creates and the reductionisms artifacts foster that are shared by both scientists and their critics; and instead of rejecting evidence that scientific methods might be culturally shaped, we need to wrestle with how that cultural influence has constrained the knowledge that could be gained. As a result, the standardization of time in science has led to representing the many cycles of living organisms in terms of artifactually mediated time, not in relationship to the Earth's somewhat variable rotation and cycles of daylight. Precision of measurement with devices that conflate timing and duration in the study of biological cycles has overshadowed the understanding of the relationship between the solar cycles and biological cycles. In a strange, ironic way, then, the imposition of artifactually mediated standardized timekeeping on the study of biological rhythms has generated a body of knowledge on the relationship between these rhythms and the homogenized, standardized time created in processes of globalization, but not a body of knowledge on biological rhythms in relationship to the Earth's erratic cycles of rotation and daylight. Both science and social science have metamorphosed globeness into absolute, globally transcendent concepts of time.

Linking natural and social phenomena is crucial to understanding social time (Adam 1995, 51; Elias 1992, 86), and consequently to understanding the interaction of de-globed, homogenized time to local experiences of multiple socially and environmentally embedded cycles. The natural sciences' adoption of de-globed artifactual time combined with the social sciences' eschewing of biological knowledge obscures the ability to understand not only the interaction of nature and culture, but also of the local and global.

Embodied Contradictions

Chronobiologists recognize that humans, while diurnal animals, are unusual among animals in their ability to exert "volitional control over their temporal niche (e.g., in the form of self-imposed sleep deprivation, night work, and high-speed travel across multiple time zones)" (Dijk and Edgar 1999, 112). Consequently, when studying humans in their natural habitat, as opposed to in the laboratory, it is not possible to understand many human biological cycles without reference to humans' choices and social lives. Even in the laboratory, in discussing a series of experiments on humans in isolation, Wever notes that, in humans, social contact is a powerful zeitgeber (1979, 151). McEachron and Schull point out that such entrainment does not mean that humans' biological diurnal tendencies are overcome, but that instead, current social relations, particularly in the workplace, create pressure "to be active behaviorally at times when our endocrine systems are most urgently demanding retreat" (McEachron and Schull 1993, 336). There have been multiple studies of the effects of artificial light on the human biological clock, and all demonstrate that this clock is influenced by such cues (Boivin and Czeisler 1998; Koller et al. 1994; Trinder et al. 1996; Waterhouse et al. 1998; Wehr 2001; Wehr et al. 1995; Wright et al. 2005; Zeitzer et al. 2000). One study even noted that the timing of melatonin secretion shifted by one hour whenever subjects changed between daylight saving time and standard time (Wehr et al. 1995, R177). Consequently, any social activity, either labor or leisure, that relies on artificial light at night will have an effect on human biological circadian cycles. One chronobiologist mentioned "prime-time TV scheduling" with early start times for morning shifts as two factors that can disrupt sleep (Monk 2000, 90).

There are particular hormones that are implicated, of which I shall here emphasize cortisol, thyrotropin, and melatonin. Importantly for understanding the experience of one's body, these hormones are also implicated in metabolism, as well as in many mental states and psychiatric disorders, particularly mental states related to stress. Cortisol plays a major role in protein and lipid metabolism and increasing blood glucose levels, and it is also part of the body's response to stress. In diurnally active humans, cortisol rises just before awakening and then begins to decline during the first hours of wakefulness. Thyrotropin works on the thyroid to generate the production of thyroid hormone—a hormone involved in cell metabolism. Its levels peak at night and then decline during daylight periods. Melatonin acts to promote sleep at night and peaks before the onset of sleep. In sum, metabolism, and consequently the physiological basis for physical activity, is tied to temporally sensitive hormones. Among diurnal primates (including humans) the cycles of these hormones are synchronized. Among humans that deviate from diurnal activity cycles, the levels and cycles of these hormones change, and their relationship desynchronizes. This is most acutely felt by most people the second or third day after a significant time shift, such as that caused by jet lag. The reason for this is that on the first day all the hormone rhythms are equally out of sync with the environment, but they adjust to the new temporal setting at different rates (Foster and Kreitzman 2004, 203).

One of the elements of Foucault's insights on discipline (1977) is the importance of inducing stress as a means of achieving discipline, and one strategy for inducing such stress is temporal control. Physical and emotional stress can result in elevated cortisol levels, and temporal discipline, if it disrupts sleep/wake cycles, also affects cortisol, thyrotropin, and melatonin levels. Foucault's emphasis on embodiment does not invoke biology, but with regard to circadian cycles what Foucault argues is consistent with what endocrinology suggests—that temporally disciplined activity should elevate and alter thyrotropin and cortisol cycles, and if it occurs at night, should suppress or at least delay melatonin secretion. Foucault's discussion does invoke cognitive artifacts—clocks, bells, and drums. All of these are not just cognitive artifacts but also tools of power. Their use as a means to know the time is tied to their use as a means to discipline and coordinate action according to the time they define.

The research on night-shift work demonstrates this flexibility in individual hormone levels (Weibel and Brandenberger 1998), but also indicates many negative consequences of the desynchronization of hormonal cycles that results from such work. These consequences include an inability to get sleep comparable to nocturnal sleep, increased stress, diminished cognitive performance, and negative health effects (Knutsson 2003; Rajaratnam and Arendt 2001). Such desynchronization importantly demonstrates that it is possible to create circumstances in which the cycles of these different hormones are not directly linked, but can become unhealthily independent—night-shift work disrupts the normal relationship of hormones, a state known as internal desynchronization. In night-shift workers, cortisol remains at higher levels throughout the period of sleep than is the case with day-shift workers, but during working hours is lower for night workers than for day workers, and it declines throughout the working period. Thyrotropin is lower during sleep for night workers than for day workers, but during a worker's shift, for night workers thyrotropin levels begin lower than those of day workers, but slope upward and are higher than day worker levels at the end of an eight-hour shift—its slope moves in opposite directions during activity for day and night workers (Weibel and Brandenberger 1998). Nighttime physical activity influences circadian cycles (Baehr et al. 2003; Mistlberger and Skene 2005; Mrosovsky 1996; Mrosovsky et al. 1989). It increases cortisol levels and causes phase delays in melatonin secretion (Buxton et al. 2003; Monteleone et al. 1992; Van Reeth et al. 1994), and causes a delay in the phase of temperature rhythms (Eastman et al. 1995). Nocturnal exercise also has an effect on thyrotropin levels, with early-night activity increasing nocturnal levels, and predawn activity delaying the normal morning reduction of thyrotropin levels (Van Reeth et al. 1994). These results are consistent with the growing awareness that physical activity can affect the phases of circadian cycles (Dijk and Edgar 1999, 115; Mrosovsky 1996). Indoor lighting is sufficiently strong to affect circadian cortisol and melatonin rhythms (Boivin and Czeisler 1998). These hormones, then, all demonstrate differences between night workers and day workers, but these hormones do not show the same changes. It is not possible to conclude that night work breaks down biological rhythms—instead, it fragments the coordination of the body's multiple hormonal rhythms.

The 24-hour readiness demanded of time-space compression can have a biological signature of internal desynchronization; the social arrhythmias encouraged by the use of cognitive artifacts have biological consequences. These biosocial arrhythmias gain significance by their disruptions, and disruptions that beget further disruptions, for example, the overly tired truck driver who is involved in a crash shutting down a major highway just before the morning commute begins, or even worse, the nuclear accident in the early morning hours that causes a mass evacuation of a city.

This has health consequences. There is a large literature on the effects on productivity and health of workers laboring at night. In fact, night-shift work is a domain where the flexibility of human cultural behavior and the artifactual representation of homogeneous time confronts inflexibility, temporal heterogeneity, and timing in human biology. Weibel and Brandenberger discovered that night workers who were satisfied with their schedule still showed signs of internal desynchronization after two years of night-shift work (1998, 206). In night workers, cortisol levels were unusually elevated and thyrotropin levels were unusually low when they were sleeping during the day. In comparing hormone levels at work, night workers had cortisol levels that were lower than their daytime counterparts, but still higher than when their daytime counterparts were sleeping. In effect, night workers had consistently elevated levels of cortisol. The difference between the cycles of different hormones is striking. Thyrotropin, which influences cell metabolism, ceases to manifest a cycle that acts in concert with cortisol, a hormone that affects body metabolism, but instead cycles independently of cortisol under such circumstances. As Foucault implies, cognitive artifacts can be used to entrain human activity, but Foucault does not address how such artifacts do not have absolute power over human biology. In the case of entrainment, resistance can be biologically embodied.

Many diseases that are related to stress—for example, heart disease, high blood pressure, and cancer (see Sapolsky 1998)—and the disruption of embodied, endogenous circadian cycles increasingly replace contagious diseases, workplace accidents, and malnutrition as major causes of worker mortality. Whitehead, Thomas, and Slapper associate peptic ulcers, cardiovascular mortality, chronic fatigue, excessive sleepiness, difficult sleeping, increased divorce rates, increased rates of substance abuse, and depression with night-shift work. The study commissioned by the US Congress found that night-shift workers' desynchronization with daily social cycles disrupts their ability to meet responsibilities, and it creates a feeling of alienation from the community because of the difficulty night-shift workers have in participating in local recreational, social, and religious events (US Congress, Office of Technology Assessment 1991). The *Diagnostic and Statistical Manual IV-TR*, the standard psychiatric diagnostic reference issued by the American Psychiatric Association, has a category for "Circadian Rhythm Sleep Disorder," which is caused by "a mismatch between the individual's endogenous circadian sleep-wake system and exogenous demands regarding the timing and duration of sleep" (2000, 622). This reference describes the symptoms of this disorder as "clinically significant distress or impairment in social, occupational, or other important areas of functioning" (2000, 629).

These deleterious effects are not only caused by night-shift work; long-term restrictions on sleep also have a negative effect on alertness and mood (Czeisler 2003; Dinges et al. 1997; Van Dongen et al. 2003). Folkard and Barton (1993) demonstrated that work shifts that start early in the morning, for example, at 7:00 AM, can disrupt sleep in ways similar to night shifts. One study found that people who chronically get six hours or less of sleep per night show cognitive impairments similar to people who have had two successive nights without sleep (Van Dongen et al. 2003), which means that the extension of the workday into the early morning and evening and the emerging importance of flexible schedules (Costa et al. 2004; Presser 1999; Sennett 1998) can influence human circadian cycles. When this flexibility is demanded by the employer, rather than chosen by the worker, it tends to lead to negative effects on sleep and mental health, and increases stress (Costa et al. 2004; Janssen and Nachreiner 2004), with flexible schedules that disrupt circadian cycles having the greatest negative effects (Giebel et al. 2004). Shift work, then, is not the only activity pattern with chronobiological consequences. Many patterns of labor and leisure that result from the use of indoor lighting and the nocturnal use of telecommunications that link the globe can restrict sleep in ways that, over time, share some consequences with shift work.

The problems associated with sleep deprivation are also associated with sleep restriction—obtaining less than seven hours of sleep each night. Sleep restriction has been associated with increased risk of mortality (Dinges et al. 2005; Gallicchio and Kalesan 2008), increased likelihood of obesity (Magee et al. 2008), increased probability of cardiovascular incidents (Rüger and Scheer 2009), decreased alertness and increased rates of on-the-job mistakes and accidents (Dinges et al. 2005), reduced psychomotor vigilance (Peters et al. 2009), decreased academic performance among preadolescents (Meijer 2008), diminished response time on cognitive tasks in women (Stenuit and Kerkhofs 2008), and suppressed hippocampal neuron creation with negative consequences for memory function (Meerlo et al. 2008). Such patchwork indicators of negative physical and mental consequences of sleep restriction have led the US Institute of Medicine and others to suggest that the study of the consequences of sleep in response to changing work and activities patterns is understudied and not sufficiently understood (Institute of Medicine 2006).

The problems created by sleep restriction and sleep deprivation not only affect the laborers involved, but have potentially broader consequences. Moore-Ede (1993) notes that many major disasters caused by human error—the Exxon Valdez oil spill, the nuclear accidents at Three Mile Island and Chernobyl, and the Union Carbide disaster in India—all occurred at night, and were associated with excessive overtime and poor shift-work scheduling.

The pressure of artifactually mediated time-space compression not only can cause internal biological desynchronization, but can also disrupt the relationship between the body's processes and its environment—or external desynchronization. Social arrhythmias have physiological consequences. Athletic competitions are particular examples of events tied to a particular locale in which pharmacological intervention is not possible (indeed, it is prohibited). During the 2004 Summer Olympics in Athens, many journalists and commentators asked why events were scheduled during the heat of the day. For instance, both the men's and women's marathons, which included many hills, were scheduled at 6:00 PM local time, whereas in the Sydney Olympics in 2000, such events were scheduled for the early-morning hours. To make matters worse, the endogenous cycle of core body temperature in humans peaks around the early evening hours. Consequently, the heat of the time of day interacted with the cycle of body temperature to exacerbate heat-related problems. Both leading up to the race and during the race the heat of the day during the early-evening hours was noted, particularly when the British runner Paula Radcliffe dropped out of the women's marathon due to heat exhaustion.

The scheduling of the marathon events was not the result of local tradition or conditions. The previous October, another marathon had been run in Athens, and it had begun at 8:30 in the morning. Throughout the world, the typical start times for competitive marathons is in the morning. The scheduling for the Olympics was anomalous for marathon running and seemed driven by maximizing the audience that inspired the greatest advertising revenues, namely, Western Europe and North America. In effect, knowledge of clock time half a world away determines the timing of events.

The Masking of Embodied Contradictions

Internal desynchronization and the health consequences of this are part of the physical costs of labor generated by new global temporalities structured by objects of time. The physiological contradiction is not presently resolvable except by using telecommunications to have daytime workers available to nocturnal interests half a globe away. The physiological contradiction can be masked, however, by a combination of intense social activity, intense physical activity, and pharmacological interventions such as stimulants to keep one awake and sedatives to force sleep during daytime hours. The clock-assisted scheduling of the timing and intensity of light has become an important component in strategies to assist night-shift workers and those who suffer from jet lag to adapt to their schedules (Boivin and James 2002; Bonnefond et al. 2004; Dumont, Benhaberou-Brun, and Paquet 2001; Eastman and Martin 1999; Monk 2000; Revell and Eastman 2005). But at base, this substitution of lights triggered by clock time for the solar day is not completely successful as the data on night workers suggest—their bodies are still internally desynchronized. The homogeneous logic that undergirds clock time cannot fully erase and decouple the ties of the human body to cycles of daylight.

Even as bodies and social networks are made arrhythmic and desynchronized, technologies are deployed to create an image of uniformity. Now, however, it is not an elision of the difference between the clock and the calendar, on the one hand, and the irregularity of the Earth's rotation and revolution, but on the other, an attempted suppression of knowledge of embodied desynchronizaiton. Not only is capitalism attempting to reduce circulation time to zero, but since humans are still a crucial link in capitalist circulation, there is a growing logic that wishes the body could be made to be permanently awake, for then the arrhythmia would disappear.

So drug "therapies" are being devised to attempt to homogenize the body's cycle to be consistent with homogeneous time, with modafinil (Provigil) being the most prominent example. The emergence of this drug is not surprising: as Foster and Kreitzman note, circadian-related disorders cost an estimated 40 billion dollars in the United States in effects on production, use of medications, and increased accident rates. They add, "it is a virtual certainty that new, more targeted, pharmaceutical interventions for sleep disorders will become available" (2004, 198). Modafinil, which is hailed by some as a miracle drug that allows people to stay awake without any sign of the health consequences of amphetamine abuse, seems ideally designed for the challenges the body faces under conditions of time-space compression. While these uses are increasing, the approved use of the medication is for narcolepsy only, although most prescriptions are for uses other than those approved for the drug, including the treatment of depression, fatigue, and as "go-pills" for US Air Force pilots on long missions (Barrett 2004). It has received increasing attention as a means of treating shift-work sleep disorder (Czeisler et al. 2003; Revell and Eastman 2005; Walsh et al. 2004), although research that evaluates the alertness of night workers on the drug throughout their shifts finds that these workers still were susceptible to severe sleepiness and diminished psychomotor performance at night (Czeisler et al. 2003, 2005). In addition, modafinil does not have an identical effect in all cognitive tasks (Turner et al. 2003), with tasks of short duration showing the smallest difference between subjects taking modafinil and those taking a placebo (Walsh et al. 2004).

Not surprisingly, media claims are exaggerated. As a report in the *New York Times* said, "In a culture of 24-hour stores, graveyard shifts and coffee shops on every corner, modafinil might also pose a more subtle danger: to the countless Americans in search of an extra edge, modafinil could be a cure for sleep" (O'Connor 2004, F1)—a "cure" for sleep indicates that some, at least, view the body's need for sleep as pathological. Such reports also indicate a desire by some for such a cure, which is echoed by scientists who document the limitations of modafinil—"the residual sleepiness that was observed in the treated patients underscores the need for the development of interventions that are even more effective" (Czeisler et al. 2005, 476). So even the science that demonstrates the "stability" and "precision" of endogenous circadian human biological rhythms (Czeisler et al. 1999) desires therapies that manipulate these cycles to promote nocturnal alertness, diminish the health consequences of night work and sleep deprivation, and basically suppress biology. Yet, even as this desire for curing the need for sleep is expressed, some offer a warning that the use of modafinil "is not natural and may cause ethical attention whenever it is regularly used in work situations, for its long-term effects after several months or years are still today completely unknown" (Bonnefond et al. 2004).

So the globeness of time-space compression not only creates internal biological conflicts, but is met with new therapeutic interventions. The desire for such therapies drives consumers and health professionals to find more effective treatments for the pathological consequences of forcing humans to work at night or with too little sleep. Will this type of commodity become one of the new basic subsistence needs? Will the attempt to create the illusion of the uniformity of time across the globe overshadow the desynchronization of the body and social life? Will the new tools of time involve a never-ending diet of stimulants? How much can the temporal uniformity of modern timekeeping artifacts reprogram the body and its rhythms?

Global Distributions of Power and Embodied Contradictions

The global distribution of these embodied contradictions and the technologies and the drugs used to address them are not even. Likewise, the distribution of the power to influence how time-space compression is experienced in a particular time zone is not evenly distributed, either. The technologies, drugs, and power are concentrated in Europe, East Asia, and North America.

The motivation for time-space compression is often described in terms of taking advantage of technology that abbreviates or eliminates the time of transactions in order to make more money. Basically, the relationship of this idea to global capitalism is similar to the relationship of commoditized clock time to industrial production. There is one important difference, however: it is not necessary for laborers to adopt an ideology of time-space compression for the system to work—it is an ideology of those engaged in financial markets and service economies. This makes it different from the shifting temporal consciousness of workers during the Industrial Revolution. These workers adopted an equation of time and wages. In fact, the industrial capitalist link between average labor time and wage uneasily coexists with time-space compression, particularly if night-shift workers receive higher wages. One alternative to this disruption of socially average labor costs is to take advantage of modern telecommunications technology to distribute workers across the globe so that night wages are not necessary—a process noted by Friedman (2005) in his efforts to argue that the world is flat. This strategy defers to human diurnal circadian cycles rather than creating a schedule at odds with these cycles. It also suggests that some corporations recognize that the temporal qualities of a place might be an asset. So in an ideology that uses temporal artifacts to homogenize time, there are strategies that defy that logic and emphasize timing.

When viewed in these ways, it is the classes involved in flexible accumulation that are most subject to the ideology of time-space compression. This is not only implied by work habits, but also by leisure activities. As Chatterton and Hollands describe for Great Britain, "The dominant audiences of nightlife spaces are mainstream, higher-spending consumption groups such as young professionals, aspiring 'townies,' and students" (2002, 112). This creates potential splits within this class between those who reside and operate out of cities that define time, and consequently define the timing of financial markets, versus those who reside and operate out of other parts of the globe. Live events, whether they are the operation of exchange floors or sports competitions, tend to occur in greater concentrations in time-defining locations. If one takes the global cities listed by Sassen (1998)—New York,

Los Angeles, Paris, London, and Tokyo—the uneven longitudinal distribution of these cities becomes apparent, as does the fact that a huge proportion of the world's population exists outside of these cities' time zones. These cities organize the rhythms of the global economy and media. As Sassen further writes, "The ascendancy of finance and the dematerialization of many economic activities assume their full meaning only when we juxtapose the uneven temporalities they foster—juxtapositions that illustrate, for example, the disjuncture of digital and material temporalities" (2000, 222). She then adds, "Temporal features of finance capital empower it to subject other forms of capital to its rhythms" (2000, 222). Time-space compression and its associated physiological conflicts are most manifested in relationships with these cities, and consequently are unevenly distributed across classes and around the globe. In addition, time difference becomes an important display and means of reproducing the power of these locales. There are about a half dozen time zones that entrain the world.

Such differences depend not only on global telecommunications, but also on cognitive artifacts used to reckon time. These global cities also, in subtle ways, define chronobiological knowledge. The use of standardized 24-hour clock time for representing circadian rhythms is most methodologically sound for discussing city-dwelling humans who live their lives by artificial light tied to clock-driven schedules—in other words, to discuss the temporal habits of the metropolitan populations is embedded globally in the artifacts used to think about time. In such cases, the time that is faithfully reproduced by the artifacts over great distances and across national boundaries is the time of one of these global cities, not of one's location. The artifact is used to know time elsewhere, because elsewhere is more important than where one is.

Whatever one's local time, consumption patterns that rely on global communications and media tend to be in terms of the time zone of one of these global cities. Through consumption, differential power relations that privilege certain time zones can create the internal desynchronization of endocrine rhythms in people in other time zones through the external desynchronization caused by relationships that span time zones. The embodied consequences of globalization are products of both labor and leisure.

Conclusions

Time-space compression is a significant feature of contemporary global capitalism, and as such, it generates several contradictions that are then ideologically concealed. Time-space compression relies on pitting biology against society—a contradiction most acutely felt in night-shift work and jet lag. This division of nature from society is reinforced by the refusal to address the biological consequences of ignoring circadian cycles and the maintenance of a divide between biological and social science. Time-space compression generates contradictions between different local times.

Cognitively, such representations conflate two very different questions—how long a task takes versus the moment by which a task should be completed. The illusion of simultaneity and the conflation of measure and time reckoning are essential assumptions in modern cultural models of time. This leads thinkers like Glissant to attend to mundane events such as the timing of examinations in Martinique: "Candidates in an official examination (for entry into the police force, May 1979) sat their tests at *3:00 A.M.*, in order to coincide with the time of the exam in France. Imagine the candidate, driven to this examination by unemployment and who gets there by sticking close to the walls of the sleeping town, because of the risk of being arrested by a police patrol" (1989 [1981], 57–58n).

This is the case of a cognitive artifact allowing one to coordinate activities between different places in different time zones.

The absurd situation documented by Glissant can be transformed into a question: Were Martiniquan candidates taking the exam at the same time as French candidates? The simultaneity of the exams assumes that the answer is "yes"—and in a sense, the different groups of candidates were taking the exam at the same *moment*. On the other hand, 3:00 AM is not the same time as 8:00 AM. The term "time" cannot have a singular meaning, then, and its multiple senses allow for a great deal of equivocation and confusion. The multiple senses of time that can be embodied in the cognitive artifacts used to think about time leads back to several of the issues raised in the previous chapter. These devices are examples of mediated and distributed cognition. But not all senses of time are equal—those embedded in objects represent choices and omissions. The case of coordinating exams across several time zones also indicates that these devices not only mediate knowledge, but in so doing, they mediate power through both the choices of what ideas are designed into objects, as well as the choices about how those objects are used to privilege some temporal ideas over others.

The absurdity of Martiniquan students taking an examination at three in the morning is evidence of this. The logic of many of those enmeshed in globalization is that certain activities must take place simultaneously across the globe, even though such simultaneity is also an illusory product of modern concepts of time (Galison 2003). This logic of simultaneity contains the often unacknowledged consequence of the time of convenience in the metropole determining the time of execution of activity globally, regardless of the time differences.

Time-space compression generates conflict between production and labor. Lacking biology and a body, capital accumulation defers to no biological clock but instead to the artifacts to which it gave birth, but the experience of human labor and consumption includes the effects of the timing of these processes on the body and its cycles. The consequence of disrupting human circadian cycles in favor of temporal uniformity is both internal and external desynchronization. This creates short-term problems in mood, disrupted sleep/wake cycles, and cognitive performance that are addressed with drugs, and it substantially increases the likelihood of long-term health problems.

Immediate media flows of information reinforce the power of those in particular locations in controlling such flows, not only through access to technology but through timing. To access immediate information, much of the world must realign its temporal habits to a very limited number of influential places. Consequently, those who control the timing of the distribution of important information are able to affect the body habits and physiology functions of all those throughout the world who rely on such information. This effect is not complete control, however, but instead creates conflicts between different physiological cycles that are adaptive when synchronized. Time-space compression does not make a flat Earth, but by virtue of existing on a globe, makes some of capitalism's contradictions physically and mentally excruciating for those touched by such compression.

So we arrive at an important irony, that the so-called emphasis on space found in recent social theory results in globalization without a globe. This is made possible by temporal artifacts with logics of uniformity that erase the cycles of our globe. Consequently, it has an inadequate means of capturing the material, environmental, physical, and biological influences on human activity, the technological and pharmacological attempts to overcome those limitations, or the new means of temporal exploitation. This new temporal exploitation is not based simply on the quantity of labor or on the productivity of labor, but on the timing of labor coordinated by artifacts and sometimes at odds with the location of the labor. The power of cognitive artifacts to channel cognition becomes the power to foster and even demand the synchronization of activity around the globe to a few locations. The faithful reproduction of the logic of determining time results in the faithful reproduction of the time of a half dozen global cities. The results are social and biological arrhythmias.

Chapter 6

Creeping Cognitive Homochronicity and the End of the Time of Earth

There is no need to turn toward the mind or subjectivity to escape from cold and objective time to find the rich "lived" world of meaning. To find richness, one only has to turn toward the world itself, to the wind, the foam, the sun, the snow-capped mountains in the background, the earnest miniature city behind the harbor. "Objective" time and "subjective" time are like taxes exacted from what people live in the world, they are not all that these multitudes do and see and mean and want.

—*Bruno Latour (1997, 172)*

In the past, major changes to systems of time reckoning were noted. Now, objects of time so powerfully mediate cognition and obscure the logics of time that fundamental changes to how time is defined go unnoticed.

When Julius Caesar reformed the Roman calendar, he not only named the seventh month after himself, but the reforms also involved creating a year of 445 days for one year. This elongated year was called "the last year of confusion" by Macrobius (2011, 167 [fifth century, 1.14.3]). That this was recognized as an act of power is preserved in Plutarch's recounting of a joke by Cicero that referred to Caesar's reforms: "Someone said that tomorrow the constellation Lyra would rise; 'yes', he [Cicero] said, 'on instructions'—as if this too was an imposition that mortals had to accept" (2011, 120 [first century, 59.6]). The Gregorian reforms in 1582 involved a ten-day shift in calendar dates, and in the prolonged debate leading up to Great Britain's adoption of this calendar in 1752, there was extensive public comment and great concern over how it would affect parish festivals in relationship to agricultural cycles (Poole 1998).

The gradual adoption of mean time also was publicly noted. In eighteenth-century Berlin, the clock of the Berlin Academy displayed both solar and mean times. As Sauter describes what ensued: "Thrust into public life in eighteenth-century Berlin, the Academy clock's four arms and two times were a disaster. The public could not distinguish between the arms, which meant that they did not know the " 'correct time,' and the complaints mounted. In response, in November 1787, the government ordered that the mean-time arms be removed… " (2007, 692). The solar-time arms were the ones chosen to remain—this was because at the end of the eighteenth century, Berlin still organized its day around "true time" (2007, 691). A similar conflict is described by O'Malley (1990, 1–3) with regard to the Yale College clock and the New Haven town clock at the beginning of the nineteenth century. The college clock chimed the solar hours and the town clock chimed mean hours. The discrepancy was noted by the people of New Haven. Corbin documents that such conflicts persisted in rural France until the early twentieth century (1998, 114).

Since the eighteenth century, there has slowly emerged a globalized system of uniform timekeeping. Beginning with the railroads, there has been a growing need to coordinate activities in time across large distances. This had led to elaborate efforts at standardized public timekeeping (Bartky 2000; Galison 2003) and national laws and international agreements seeking to put in place systems of temporal uniformity (Bartky 2007; Howse 1980).

Since 1967, there have been a set of changes in the measure of time that have gone without public notice. These changes fundamentally alter how time is defined. They free the measure of time from the rotation of the Earth. Since the measure is shifting away from the measure of one cycle, the Earth's rotation, and toward the counting of atomic cycles, namely, those of cesium atoms, the standard of time measurement is increasingly self-referential: it is not a single atomic cycle that defines the second, but an institutionally recognized coordination of cycles. The fundamental standard of time is defined, not measured. This standard is the product of an intergovernmental institution, and it can only be changed by international agreement. Gone are the days of Chaucer's host who calculated local time by the length of shadows and the position of the Sun. Now, objects of time mediate between the person who wants to know the time and an international administrative structure.

The central role in this structure is played by the Bureau International des Poids et Mesures (BIPM). This agency maintains all the standards for weights and measures, and at this point, time is critical because even space is defined in terms of time, with the meter being the distance light travels in a vacuum at 1/2,999,792,458 of a second. The measurement of the second is determined by atomic clocks spread throughout the globe. These signal the BIPM, which then uses them to construct coordinated time. Coordinated time is a mathematical construct, not the measure of a specific phenomenon. The reason for this is how relativity affects the relationship of space and time—there is no universal time, but only time relative to specific points in space, and this time is influenced by motion. By receiving signals from multiple atomic clocks, the BIPM is able to create a representation of time that can be globally applied even as, technically, it applies to no place. Every five days, the BIPM issues its standard time signal, which is known as Coordinated Universal Time (UTC), and monthly it publishes *Circular T,* which describes the time scales for both atomic time and coordinated time. Many time service providers, such as telecommunications companies, then use this information to manage their own time systems (McCarthy and Seidelmann 2009, 217–218).

Since the Earth wobbles, its rotation is slightly irregular. Until now, to keep UTC in coordination with the rotation of the Earth, the BIPM has relied upon the International Earth Rotation and Reference Systems Service (IERS) to advise on the insertion of leap seconds to keep UTC in sync with the Earth's rotation.

This arrangement between the IERS and the BIPM is currently defined by the International Telecommunication Union's Radiocommunication Sector (ITU-R). Within the ITU-R is a working party charged with making recommendations concerned with the management of the international time-signaling system—this is the ITU-R WP 7A.

Since 2000, Working Party 7A has been considering whether or not to eliminate the practice of adding leap seconds to UTC.

In 2011, a survey was sent to the 192 member states of the ITU-R asking if they would support a recommendation to eliminate leap seconds. The consequence would produce a difference between rotational time and UTC of about one hour every 550 years (ITU-R 2011). The battle lines in the debate are between those who see a value in a continuous time standard and those who see value in keeping an astronomical time standard even if it means inserting leap seconds. A continuous time scale is particularly useful for telecommunications and navigation, whereas a time scale linked to Earth is important in astronomy and is also the source of legal definitions of time in current precedents (Finkleman et al. 2011, 319). For most people, for whom clocks and watches are tied to time-signal providers who receive signals from BIPM, the debate is so mediated by objects of time as to be irrelevant. They are not conscious that the continuous time standard not only redefines clock time but also calendar time, since the definition of the day is now tied to the clock.

In 2012 the ITU-R debated about whether to eliminate leap seconds. With that elimination, setting time to the rotation of the Earth time would come to an end and be replaced by an atomic standard free of Earth's rotational foibles. The ITU-R decided to defer this decision until 2015.

The foundational unit of time for the BIPM is the Système International d'Unités (the SI) second. All time is currently defined in terms of multiples or fractions of this second. Even the calendar day consists of 24 hours made up of these seconds, and not fractions of the Earth's rotation. This second is defined as equal to the ephemeris second. The ephemeris second was developed as an accurate measure of astronomical phenomena, and in 1960, the 11th Conférence Générale des Poids et Mesures defined this second as 1/315,56,925.975 of the length of the tropical year for January 0, 1900 (in effect, December 31, 1899). This definition was based on the work of Simon Newcomb (1895). The tropical year is the duration of time it takes for the Sun to return to the same point in relationship to the Earth and stars, that is, the same point on the Sun's ecliptic.

While the SI second is defined as equal to the ephemeris second, since the ephemeris second is defined to a point in time now over a century past, it is not determined by current observation of the Sun in relationship to its ecliptic. Now, an ephemeris second, and consequently the SI second, is determined to be 9,192,631,770 ± 20 HZ derived from the transitions in cesium atoms. Essen (2000), one of the inventors of the atomic clock, describes these transitions as follows: In atoms of alkali metals, "The outer electron and nucleus were spinning in either the same or opposite direction and that the two conditions represented states of slightly different energies. Transitions between them were accompanied by emissions or absorptions of a (quantum of) radiation." These cyclic emissions in the microwave spectrum could be counted.

The determination that 9,192,631,770 of these cycles in an atomic clock was equivalent to an ephemeris second was made by Markowitz (Markowitz et al. 1958), but as Leschiutta observed, "it is almost impossible to explain the accuracy of the Markowitz determination" (2005, S14).

This is the standard for our clocks and calendars, since calendar days are now defined as a quantity of SI seconds. Therefore, objects of time not only mediate cognition, but also the relationship of every clock and Gregorian calendar user to this bureaucracy and the scientists who work with it. This bureaucracy can make decisions that affect what objects of time indicate without the awareness of the users of those objects.

The logics embedded in clocks and the Gregorian calendar are now global and are applied in almost all endeavors to create knowledge, and even classroom pedagogies are shaped by clock-structured and calendrically outlined lesson plans and time budgets. There is little reflection on how the artifacts' embedded logics mobilize and channel ideas of time, and how this overshadows local temporalities and specific rhythms. Moreover, these artifacts preconceive the articulation of the local and the global. In a brute sense, there are local solar times and local biological and environmental cycles, but these become represented in terms of clock and calendrical time that are becoming disengaged from biological cycles and the Earth. In effect, the current generation of temporal artifacts hides local timescapes in favor of mediating between the user of the artifact and an administrative system that maintains global standards. This makes the clock and the calendar quite different from sundials, sea-worms, or immortelle trees, all of which are extensions of the mind that focus attention on the local timescape rather than on standards divorced from it.

It is not only the disengagement of modern objects of time from the local environment that makes them culturally unique, but the time to which they refer. Isaac Newton argued that the time we observe in the rhythm of day and night is too irregular for science: "For the natural days are truly unequal, though they are commonly considered as equal, and used for a measure of time; astronomers correct this inequality that they may measure the celestial motions by a more accurate time" (1934 [1687], 7–8). In other words, Newton suggested that the measure of true time must be divorced from the measure of natural rhythms. Since Einstein and Poincaré, this Newtonian view has been discarded by physicists (Galison 2003), but is unquestioned in public policy, and the study of biology, psychology, culture, and society. It is the Newtonian desire for an absolute standard that is behind the ITU-R's desire to set time standards free from the Earth's irregularities.

Hassan and Purser write, "It is impossible to carry this temporal complexity in all its growing fullness into practical life; we can't think about this stuff all the time, in other words" (2007, 13). This impossibility is a feature of modernity. As I argued in chapters two and three, without the clocks and calendars, temporal complexity was part of life, and it enriched social life. People used many sources of information to determine the time, but in these cases, time reckoning was extremely contextual and the logics involved were easily learned. Referring back to the example of the parrots flying overhead in Trinidad discussed in chapter two, the parrots are only useful in recognizing a single moment in the day, not for determining time throughout the day. On the one hand, the connection between time reckoning and specific contexts creates a tight integration of time cycles and those contexts, but on the other hand, it does not allow for temporalities that span multiple contexts, much less temporal rubrics that can be applied across long spans of time, large distances, and social differences. It is not that humans lack the cognitive capacity for thinking about temporal complexity, but that temporal complexity is detrimental to the coordination and management of groups of people across contexts and space. Generalized, standardized time systems seem to be symptoms of imperialism and political centralization, and their development and maintenance require technical expertise and administrative support hidden by the objects that provide generalized and standardized time.

If one begins to look at cognitive artifacts of telling time from this perspective, then one comes to view these artifacts as channeling cognition to particular purposes rather than generally enhancing knowledge about time. The artifacts that allow for standardization across many contexts also deflect attention from contextual rhythms and cycles. Artifacts that are well suited for coordination and control over large distances and multiple situations are ill suited for adaptation to specific challenges. While this contrast is overly simplistic, it does highlight the extent to which the use of cognitive artifacts involves a choice not only about meeting cognitive challenges, but also about which challenges are addressed and neglected. Such qualities of artifacts can be acknowledged and benefit derived from doing so. For instance, Traweek, in her study of high-energy physicists, discusses how these physicists construct their own particle detectors attached to particle accelerators. In the construction, the physicists are aware of the assumptions and choices they make in the detector's design. Contrary to the standardization of clocks, the detectors vary. Traweek suggests that each detector "is the material embodiment of a research group's version of how to produce and reproduce fine physics" (1988, 72). In effect, the physicists are acutely aware that these devices mediate between the thoughts of the physicists and the world, and as a result are extremely conscious of the choices made in the devices. Awareness of how devices shape thought enrich science and spur the creation of new devices that reveal different facets of our world.

Recognition of such choices steers the narrative of the emergence of clocks and the Julian and Gregorian calendars away from a story of progress and cognitive functionalism toward a realization of the connection between imperialism, concepts of time, the artifacts that shape cognition, and the uncritical reliance on these objects. The Julian calendar derives its name from its implementer, the emperor Julius Caesar; the Gregorian reforms were enacted at the height of the Spanish Empire in the New

World; Great Britain's adoption of the "popish" calendar coincided with that empire's ascending global and economic domination. This is not merely a European pattern: imperial calendrics are associated with the Chinese dynasties, the Maurya dynasty in South Asia, the Aztecs, the Maya, the Inca, and the ancient Egyptians among others (see Aveni 2002; O'Neil 1975).

Clocks are associated with the measurement of labor time and with work-discipline in capitalism. Greenwich is the prime meridian because of English imperial and economic power, not because that suburb of London is a good place for astronomical observations. The global standardization of time zones was led first by the railroads, and then by the US government as its own imperial designs increased. The move toward atomic clocks, and their adoption as the standard of time in 1967, was dependent upon global telecommunciations and coincident with the integration of the global economic system in which minutes, even seconds, count in financial transactions. In her review of anthropological studies of time, Munn notes that "chronological instruments" like clocks and calendars, what I would call cognitive artifacts, connect everyday life to larger issues of governance and control—"It has to do with the construction of cultural governance through reaching into the body time of persons and coordinating it with values embedded in the 'world time' of a wider constructed universe of power" (1992, 109).

The move from local cycles to abstract globally implemented cycles was achieved in the calendar long before the clock. The purpose of the Julian calendar was standardization across the Roman Empire, and subsequently, throughout the Christian Church. As discussed in chapter three, the Julian calendar and its successor, the Gregorian calendar, employ standardizing logics that combine the measure of duration with the determination of points in time. They suppress other cycles, and this suppression is what posed the cognitive challenge of developing formulae to determine the calendrical date of Easter that fostered what has become known as the computus (Baker and Lapidge 1995; Borst 1993; Bradley 2002; Jones 1943). This was no simple task, yet it was a necessary one for priests to be able to successfully meet so that they would start Lent and its fasting at the appropriate time.

The calendar had already moved from a counting of time to an artifact that anachronistically and seemingly illogically preserved the count—beginning with the ninth month of the year all the month's names are misnomers: the name *Sept*ember implies that it is the seventh month. European time-reckoning history can be viewed as a transition from temporal rubrics that were localized and based on counting to an eventual rubric based on universalized (literally, since space missions follow UTC), uniform time that retains survivals of the counting-based logic.

If one attends to the measurement of time and how cognitive artifacts mediate cognition, culture, and power, then the computus and Easter tables, and not the sundial, are the precursor to the modern clock's role in mediating cognition about time. It is in these devices to calculate Easter that one observes objects used to mediate between users and a universalizing institution. The similar appearance of some sundials and clocks falsely gives the impression of a unilineal evolution from one to the other despite the overwhelming documentary evidence that they offered contrasting ideas of time not just in terms of the contrast between the canonical hours and clock hours, but also in terms of the contrast between solar time and mechanical time as described in chapter one.

This suggests that what drove the standardizing logics of clocks and calendars was not technological progress but managerial desire—the desire of large institutions to control and coordinate activity over large distances, and to make standards of duration in one location equivalent to those in another regardless of the changing amounts of daylight. The reason for thinking it is managerial desire rather than technical knowledge is that the ability to create standard durations is very old—water clocks did this. Moreover, King Alfred's candle clock is an example of a device used to measure activity that was in the hands of a monarch but did not result in widely imposed temporal standards. According to Alfred's biographer, Asser, the king desired to spend "one half his bodily effort both by day and by night" in service to God (1983 [893], 107–108). But there were two substantial challenges to Alfred's being able to meet this goal. The first challenge, which Asser notes, was that "he could not in any way accurately estimate the duration of the night hours because of darkness, nor the day-time hours because of the frequent density of rain and cloud" (Asser 1983 [893], 108). He decided to use candles—a technique that would measure equal durations. He ingeniously constructed a lantern that used slices of ox horn to shield the candle from drafts and the wind. By means of this tool, Asser reports that Alfred was able to achieve his goal and govern his kingdom. The second challenge, which Asser does not report, but which had to have been a factor, was the seasonal variations in daylight—there was no means by which one could use canonical hours that were seasonally variable and achieve the equal division of time throughout the year that Alfred sought. This is why monastic rules had seasonally specific activity cycles. So King Alfred set out to do something quite unusual—to measure all his activities using a logic previously used to measure a speech. To me, what makes Alfred's invention remarkable is not his use of a candle to tell time—that had probably been done before in Europe, and there is evidence that incense was used in this manner in Asia (see Bedini 1994). Alfred's invention is remarkable because of the choice embedded within it. For his personal devotional practices, rather than emulate monastic communities and their seasonal schedules as laid out by St. Benedict's Rule, Alfred made a different choice that was immune to seasonal variations. His choice was highly idiosyncratic, and was not imitated. Yet, his choice shows that there was the technological capability and cognitive know-how to combine the tasks of measurement and determination of points in time. His choice shows that it was a choice, and not simply a consequence of the invention of the clock.

Alfred's candle clock suggests that the dominance of the logic of clocks is not because technology created widespread desire. The form of modern time technologies is based on the standardization of human routine rather than an accurate link between technology and environmental cycles.

The Gregorian calendar reforms is another case of the cognitive tool being an object of power because it shaped thought, and not because new technologies suddenly became available. The problems with the Julian calendar were noted centuries before the calendar reform—the knowledge was there to change the calendar. Instead, the change coincides with the expansion of empire and commerce at the end of the sixteenth century. The Julian calendar, and consequently the Church's determination of the spring equinox and Easter, were known to be wrong for centuries before the Gregorian reform (North 1983). After the Gregorian reform, the inaccuracies of the new calendar were known, and were raised in debates in Protestant nations about the adoption of Pope Gregory's reform. For instance, Leibniz created a plan that the Protestant states of Germany could accept (Leibniz 1997 [1700], 807–810), namely, to adopt the Gregorian calendar, but to do so while replacing the Roman Catholic Church's tables to determine the date of Easter with astronomy. In Leibniz's letters, one can sense the conceptual division he drew between astronomical reckoning, on the one hand, and popery and popular custom, on the other: "Indeed, if Gregory had restored calendars to the true laws of the heavens, and had made astronomers their guardians, nothing could have been added to his foresight" (1997 [1700], 807). Leibniz advised the Royal Society in England (1997 [1700], 807–810), but his recommendations were greeted coolly, in part because of the growing hostility between him and Newton. By the May after Leibniz's letter, Flamsteed, the English Astronomer Royal, reported having a paper written by Newton annotated as "Mr. Newtons [*sic*] paper in Answer to it [Leibniz's letter]" (Flamsteed 1997 [1700], 823). Newton was not merely content to criticize Leibniz's adoption of the Gregorian calendar for secular purposes, but proposed an astronomical solution to the calendar problem that was more astronomically accurate, albeit more complicated, than Leibniz's. Eventually, Great Britain simply adopted the Gregorian calendar, rather than Newton's complicated, yet much more accurate, proposal. The need to coordinate trade and social relationships trumped scientific desires for accuracy. This, too, was a choice. Our continued use of the Gregorian calendar despite its flaws is also a choice, but is not a choice of its users. Like the debate over the leap second, to change the calendar would involve international conferences, committees, working groups, bureaucracies, and ultimately international agreements. The pope no longer has control over the pope's calendar.

The clock and the Gregorian calendar as we know them are not some simple consequence of technological or scientific progress, then, but reflect choices that appropriated and enhanced existing technical capabilities for purposes of management and coordination. The use of objects to distribute particular logics is a feature of empire and commerce.

What began as tools now have gained the status of signifiers of natural time with the "nature," in fact, being the logic of the signifier. Like Austin's (1962) concept of a speech act, these cognitive artifacts create something through self-referential signification. Such signification does not automatically produce the clock and calendar time that we have inherited, but instead is involved in all systems of time. Evans-Pritchard's well-known discussion of the Nuer cattle-clock is an example. Nuer daily time reckoning refers to activities of caring for cattle, for example, "taking the cattle from byre to kraal, milking, driving of the adult herd to pasture, milking the goats and sheep, driving the flocks and calves to pasture, cleaning of byre and kraal, bringing home of the flocks and calves, the return of the adult herd, the evening milking, and the enclosure of the beasts in byres" (1940, 101). This list of time references contains activities that would have roughly the same duration each day, such as milking, and other events that involve a specific moment during the day, such as "the return of the adult herd" (1940, 101–102), but they are also activities in which there would have been seasonal variation, which Evans-Pritchard does not describe relative to time reckoning, although he does imply seasonal variation in activities only a few pages before. Yet, in a sense, the seasonal variation possibly mattered as little to the Nuer as the timing of sunset to six o'clock in clock time—the system is self-contained and self-referential. The difference between the two systems is that the Nuer do not use their cattle-clock to measure duration, whereas the clock is employed in this way, but this difference merely is a cultural variation built onto the basic self-referentiality of the time-reckoning system.

Even as our self-referential objects of time that emphasize temporal uniformity encourage institutionalized social coordination, they impede our ability to represent, much less to think about, other temporalities and their self-referential logics. Something that is rarely pointed out ethnographically, except for Subarctic and Arctic societies, is that there are many cultural logics in which the time of day is determined by pointing to the position of the

Sun or by the behavior of animals (e.g., the crowing of roosters), but seasonal variations in these methods of time reckoning are never mentioned. It is as if every society, with the exception of Arctic societies, is represented as if it was located in the tropics.

The logic embedded in objects of time is so naturalized that it even shapes ethnographic representation. Malinowski's (1927) use of the Gregorian calendar to represent Trobriand events indicates that such a calendar is employed even when working with people who use non-Gregorian systems of time reckoning. In dealing with the calendric intricacies and conflicts he found in Bengal, India, Klass begins to broach the problem of the temporal frames used in ethnography. In his account, he was dealing with disagreements between Bengali Hindus over what year and month it was, although "many, perhaps most, religious holidays are celebrated throughout India on exactly the same day" (1978, 167). To explain this phenomenon, Klass points out, "The era, the year, the month, and even the day of the week are all irrelevant to the calculation of the occurrence of religious events. Rather the times of religious observance are determined with the aid of zodiacal calendars found throughout India. Only two steps are involved: determining in the correct lunar period (waxing or waning of the moon) and then within that the exactly correct 'moment' in time for the ceremony" (1978, 167). He says that in response to this diversity, "the fieldworker charts, month by Indian month, the festivals and ceremonies observed in his particular village of study" and then he suggests that the product of this exercise, "the 'village calendar'...runs a strong risk of being a construct of the anthropologist" (1978, 169). Bourdieu's worry about the calendar as constructed by analysis is similar (1977 [1972], 98). Both confront the problem that the abstracting homogeneous logic of our objects of time conflicts with local cycles that emphasize timing and not homogeneous durations.

The dominance of the Gregorian calendar is not merely in ethnographic practice, but extends to non-Christian religious time reckoning. Muslims, Jews, Hindus, Chinese, and Old Calendarists of the Eastern Orthodox Church all maintain chronologies and calendars that do not fit with the Gregorian calendar, but for many of the faithful, their primary means of knowing when a holiday is in their tradition is found in a Gregorian calendar. The vast majority of interfaith calendars on the Internet are Gregorian calendars with the holidays of other faiths placed in them. In an informal poll of ethnographers who work in various parts of the world, it turns out to be quite common. My colleague Murphy Halliburton told me that in Kerala, India, the common form of calendar has a Gregorian organization but includes references to the Hindu calendar and important days in the Hindu year; another colleague, Mandana Limbert, says that the school year in Oman is represented in a Gregorian-style calendar that includes references to the Muslim calendar; and another anthropologist, Robert Canfield, confirms that this use of Greogrian calendars in Muslim countries is widespread. In Trinidad, all religious holidays are represented according to this Christian calendar, even though the Hindu and Muslim calendars are quite different from the Gregorian. Such evidence supports Postill's claim that clock and calendar times regulate the "daily rounds of most people, artifacts and representations across the world" (2002, 251).

The taken-for-granted homochronicity that the clock and the Gregorian calendar generate hides a history of heterogeneity, conflict, and struggle. The consequence of this is that any techniques or methodologies that assume clock time or Gregorian calendar time run the risk of distorting phenomena to make them fit into a post-Enlightenment European temporality that emphasizes homochronicity, and such distortion hides the physical, phenomenological, cognitive, and cultural implications of the interaction of different temporalities. Even those most attuned to cultural variation have had their ability to think about time artifactually shaped.

Chapter four's attention to how clock and Gregorian calendar time channel thought and are maladaptive for certain polyrhythmias is an indication that the functional dimensions of these cognitive artifacts have a dysfunctional cost. The evolution of modern time technology was not a functional necessity, but an act of power. Time systems exist to emphasize and privilege certain social functions at the expense of others. The channeling of thought, then, is not merely about time, but is about priorities with those activities that are most readily adapted to the artifacts of time receiving greater emphasis than those that defy those logics. The concentration of these logics and their power in artifacts, and the universality of the time produced grant power over time to human institutions and deflect attention from the intersection of environmental cycles. The experience of time, and often the relationship of time to the sacred, is no longer tied to the immersion of intersecting rhythms in a particular place, but instead is coordinated by a centralized authority.

The cognitive artifacts of time that are now globally ubiquitous—calendars and clocks—do not indicate natural time or real time. They are cultural products used to think about time. They are products of recent origin, and their widespread adoption is even more recent. Gurevich says of them, "The present perception of time bears very little resemblance to that of other epochs" (1976, 230). In effect, in thinking about the human understanding of time through the human past and across cultural differences, we have adopted a unique and artifactually mediated set of ideas as the ideal type against which all other ideas are understood and evaluated. This poses a problem when we convert other ways of thinking about time into homologues of the clock or calendar. The Nuer do not have a cattle-clock; they have a cattle-based way of reckoning time. In the Trobriands, there is no calendar with a core logic that needs to be deciphered—there is a recognition of seasonal and annual rhythms that can only be conceptualized if one allows the Moon to "go silly" once in a while. The medieval canonical hours are points in time, not standard durations.

Time is a topic where commodity fetishism, the extension of cognition outside of the mind, and the determinism of cognition meet. There is a great deal of human cognition that relies on the external world, and the measurement and determination of time is one such area. Intersubjective agreement about time depends on an intersubjective standard, often embedded in the environment or an artifact. Marx's concept that commodities mediate social relationships also serves as a means to think about how commodities can mediate cognitive relationships, as well. Those commodities that assist our thoughts do so as a result of design features in which logics are embedded. The fetishizing of the cognitive artifact obscures the relationship we have to its designer. Their identities are unknown to most, even as their logics, choices, and thoughts determine how most people think about time. It is this dual feature of cognitive artifacts as not only mediating cognition but also shaping cognition in their users that then becomes a means by which one can contemplate how powerful cognitive artifacts can be. It is easy to criticize clock time and suggest that it is artificial, but such criticism does not make it go away. It continues to shape the thoughts of the critic, if for no other reason than that it shapes the thoughts of those around the critic.

The artifacts used to think about time, then, can be viewed as an example of artifactual determinism of thought—parallel to Whorf's linguistic determinism, but unlike with language, artifacts do not have the features of being able to combine easily to create new ideas. The fixity of the artifact directly transfers to the fixity of thought. The distribution and adoption of cognitive artifacts, then, becomes the distribution and adoption of particular thought processes mediated by these artifacts. This is a distinctive feature of our species—one that separates us from our hominid and primate relatives. It is part of what makes us human.

This general feature of humans is adapted to create tools to meet specific cognitive challenges. It is also a means by which a few people do the thinking for many, but with most users of the artifacts not being aware of the thoughts or intentions of the designers. The logic placed into an artifact is often on the basis of a choice that emphasizes the performance of some cognitive tasks at the expense of others, and often, consciousness of what this choice was is hidden by the artifact itself. In this way, not only do a few end up thinking for many through the mediation of artifacts, but this mediation involves a few making a choice about how the many think. Because artifacts and their designs can endure longer than their creators, artifacts become a means to distribute and reproduce choices over large distances and many generations.

Clocks and calendars are examples of cognitive tools that enhance cognition and relieve cognitive loads, but they do so at the cost of directing cognition. This artifactually shaped cognition has become a cornerstone of the construction of knowledge. In effect, our object-mediated ideas of time are the exceptions of human history that have become the rule. Living as we do in an age in which so much of our thought is mediated and shaped by cognitive artifacts designed by people of whom we have little or no awareness and who made choices of which we are no longer conscious, we must pause to reflect on the delicate balance between tools that extend and empower the mind, and tools that constrain it.

Epilogue on the Mayan End of the World

As the supposed end of the Mayan long count approaches in 2012, I wonder if the credence given to the Mayan calendar is a result of our being so primed to uncritically trust objects of time and treat them as of universal significance.

Just as the Maya, we can treat the representations of our objects of time, particularly perceived beginnings and endings, with great significance. The recognition of the beginning of a new millennium on January 1, 2000, was an example—even though technically the millennium did not start until 2001.

Self-referential objects of time create significance that their users sometimes mistakenly attribute to the forces of the universe rather than to the self-referential features of the objects.

Bibliography

Ackerman, Robert W. 1978. "The Liturgical Day in the Ancrene Riwle." *Speculum* 53:734–744.

Adam, Barbara. 1992. "Modern Times: The Technology Connection and Its Implications for Social Theory." *Time and Society* 1:175–191.

———. 1995. *Timewatch: The Social Analysis of Time*. Cambridge: Polity Press.

———. 1998. *Timescapes of Modernity: The Environment and Invisible Hazards*. London: Routledge.

———. 2002. "The Gendered Time Politics of Globalization: Of Shadowlands and Elusive Justice." *Feminist Review* 70:3–29.

———. 2004. *Time*. Cambridge: Polity Press.

Adams-Guppy, Julie, and Andrew Guppy. 2003. "Truck Driver Fatigue Risk Assessment and Management: A Multinational Survey." *Ergonomics* 46:763–779.

Aelfric. 1875 (ca. 1006). "The Offices of the Canonical Hours." In *Select Monuments of the Doctrine and Worship of the Catholic Church in England before the Norman Conquest*, edited by E. Thomson, 113–211. London: John Russell Smith.

———. 1942 (ca. 993). *De Temporibus Anni*. London: Oxford University Press, Early English Text Society #213.

———. 1947 (ca. 1000). *Aelfric's Colloquy*, edited by G. N. Garmonsway. London: Methuen.

Åkerstedt, Torjborn, and Simon Folkard. 1995. "Validation of the S and C Components of the Three-process Model of Alertness Regulation." *Sleep* 18:1–6.

———. 1996. "Predicting Duration of Sleep from the Three-Process Model of the Regulation of Alertness." *Occupational and Environmental Medicine* 53:136–141.

Alcuin. 1844–64a (eighth–tenth century, authorship is uncertain, but attributed to Alcuin). "Disputatio Puerorum." In *Patrologia Latina, Volume 101*, edited by J.-P. Migne, columns 1097–1143. Turnhold: Typographi Brepols Editores Pontifici.

———. 1844–64b (eighth century). "De Divinis Officiis Liber." In *Patrologia Latina, Volume 101*, edited by J.-P. Migne, columns 1173–1286. Turnhold: Typographi Brepols Editores Pontifici.

———. 1994 (1895 [eighth century]). "Letter to Bishop Arno." In *Monumenta Germaniae Historica, Epistolae, Volume 4: Epistolae Karolini Aevi II*, edited by Ernst Dümmler, 319–321. Munich: Monumenta Germaniae Historica.

Alkon, Paul. 1982. "Changing the Calendar." *Eighteenth Century Life* 7:1–18.

Alliney, Guido. 2002. "Introduction to *Time and Soul in Fourteenth Century Theology: Three Questions of William of Alnwick on the Existence, the Ontological Status and the Unity of Time*," XI–XLIV. Firenze: Leo S. Olschki.

American Psychiatric Association. 2000. *Diagnostic and Statistical Manual of Mental Disorders*, fourth edition, text revision. Washington, DC: American Psychiatric Association.

Anatolius. 1926 (third century). "The Paschal Canon." In *Ante-Nicene Fathers, Volume 6*, edited by A. Cleveland Coxe, 146–151. Grand Rapids, MI: Christian Literature Publisher.

Ancrene Riwle. 1976 (late twelfth century). *The English Text of the Ancrene Riwle*, edited by A. Zettersten. Oxford: Oxford University Press.

Anderson, Benedict. 2006 (1983). *Imagined Communities: Reflections on the Origin and Spread of Nationalism*, revised edition. London: Verso.

Appadurai, Arjun. 1986. "Introduction: Commodities and the Politics of Value." In *The Social Life of Things: Commodities in Cultural Perspective*, edited by Arjun Appadurai, 3–63. Cambridge: Cambridge University Press.

Ariotti, Piero E. 1973. "Toward Absolute Time: The Undermining and Refutation of the Aristotelian Conception of Time in the Sixteenth and Seventeenth Centuries." *Annals of Science* 30:31–50.

Aristotle. 1936 (ca. 350 BC). *Physics*, translated by W. D. Ross. Oxford: Oxford University Press.

Arntz, Mary Luke. 1981. *Richard Rolle and þe Holy Boke Gratia Dei: An Edition with Commentary*. Salzburg: Institut für Anglistik und Amerikanistik, Universität Salzburg.

Asad, Talal. 2003. *Formations of the Secular: Christianity, Islam, Modernity*. Stanford: Stanford University Press.

Aschoff, Jürgen. 1965. "Circadian Rhythms in Man: A Self-Sustained Oscillator with an Inherent Frequency Underlies Human 24-hour Periodicity." *Science* 148:1427–1432.

Aschoff, Jürgen, ed. 1981. *Handbook of Behavioral Neurobiology, Volume 4: Biological Rhythms*. New York: Plenum.

Asser. 1983 (893). "Life of King Alfred." In *Alfred the Great: Asser's Life of King Alfred and Other Contemporary Sources*, translated by Simon Keynes and Michael Lapidge, 67–110. New York: Penguin.
Atkins, Keletso E. 1988. "'Kafir Time': Preindustrial Temporal Concepts and Labour Discipline in Nineteenth-Century Colonial Natal." *Journal of African History* 29:229–244.
Augustine of Hippo. 1997 (ca. 397–98). *The Confessions,* translated by Maria Boulding. New York: Vintage.
Austin, J. L. 1962. *How to Do Things With Words*, second edition. Cambridge, MA: Cambridge University Press.
Aveni, Anthony. 2002. *Empires of Time: Calendars, Clocks and Cultures*, revised edition. Boulder: University Press of Colorado.
Bachelard, Gaston. 2000 (1950). *The Dialectic of Duration*, translated by Mary McAllester Jones. Manchester: Clinamen Press.
Baehr, Erin K., Charmane I. Eastman, William Revelle, Susan H. Losee Olson, Lisa F. Wolfe, and Phyllis C. Zee. 2003. "Circadian Phase-shifting Effects of Nocturnal Exercise in Older Compared with Young Adults." *American Journal of Physiology: Regulatory, Integrative and Comparative Physiology* 284:R1542–R1550.
Baker, Peter S., and Michael Lapidge. 1995. Introduction to *Byrhtferth's Enchiridion,* xv–cxxxiii. Oxford: Early English Text Society.
Bargh, John A., and Tanya Chartrand. 1999. "The Unbearable Automaticity of Being." *American Psychologist* 54:462–479.
Barrett, Amy. 2004. "This Pep Pill is Pushing Its Luck." *Business Week* 3906:76–77.
Bartky, Ian R. 2000. *Selling the True Time: Nineteenth-Century Timekeeping in America.* Stanford: Stanford University Press.
———. 2007. *One Time Fits All: The Campaigns for Global Uniformity.* Stanford: Stanford University Press.
Becker, Gary S. 1965. "A Theory of the Allocation of Time." *The Economic Journal* 75:493–517.
Bede. 1844–64a (early eighth century). "De Natura Rerum." In *Patrologia Latina, Volume 90*, edited by J.-P. Migne, columns 187–278. Turnhold: Typographi Brepols Editores Pontifici.
———. 1844–64b (ca. 703). "De Temporibus Liber." In *Patrologia Latina, Volume 90*, edited by J.-P. Migne, columns 277–292. Turnhold: Typographi Brepols Editores Pontifici.
———. 1930 (ca. 731). *Ecclesiastical History*, translated by J. E. King. Cambridge, MA: Loeb Classical Library.
———. 1999 (ca. 725). *The Reckoning of Time*, translated by Faith Wallis. Liverpool: Liverpool University Press.

Bedini, Silvio A. 1994. *The Trail of Time: Time Measurement with Incense in East Asia*. Cambridge: Cambridge University Press.

Benedict of Nursia. 1998 (ca. 530). *Rule of St. Benedict,* translated by Timothy Frye. New York: Vintage.

Benítez-Rojo, Antonio. 1996. *The Repeating Island.* Durham: Duke University Press.

Bergson, Henri. 2001 (1913). *Time and Free Will: An Essay on the Immediate Data of Consciousness*. Mineola, NY: Dover.

———. 2005 (1911). *Creative Evolution*. New York: Cosimo Classics.

Bilfinger, Gustav. 1892. *Die Mittelalterlichen Horen und die Modernen Stunden*. Stuttgart: Verlag von W. Kohlhammer.

Birth, Kevin. 1999. *Any Time is Trinidad Time: Social Meanings and Temporal Consciousness*. Gainesville, FL: University Press of Florida.

———. 2007. "Time and the Biological Consequences of Globalization." *Current Anthropology* 48:215–236

———. 2008a. "The Creation of Coevalness and the Danger of Homochronism." *Journal of the Royal Anthropological Institute* (N.S.) 14:3–20.

———. 2008b. *Bacchanalian Sentiments: Musical Experiences and Political Counterpoints in Trinidad*. Durham: Duke University Press.

———. 2011a. "Signs and Wonders: The Uncanny *Verum* and the Anthropological Illusion." In *Echoes of the Tambaran: Masculinity, History and the Subject in the Work of Donald F. Tuzin,* edited by David Lipset and Paul Roscoe, 117–136. Canberra: Australian National University Press.

———. 2011b. "The Regular Sound of the Cock: Context-Dependent Time Reckoning in the Middle Ages." *KronoScope* 11(1–2):125–144.

———. Forthcoming. "Calendars: Representational Homogeneity and Heterogeneous Temporality." *Time and Society*.

Bloch, Marc. 1961. *Feudal Society*. Chicago: University of Chicago Press.

Bloch, Maurice. 1977. "The Past in the Present." *Man* (N.S.) 12:278–292.

Bohannan, Paul. 1953. "Concepts of Time among the Tiv of Nigeria." *Southwestern Journal of Anthropology* 9:251–262.

Boivin, Diane B., and Charles A. Czeisler. 1998. "Resetting of Circadian Melatonin and Cortisol Rhythms in Humans by Ordinary Room Light." *Neuroreport* 9:779–782.

Boivin, Diane B., and Francine O. James. 2002. "Circadian Adaptation to Night-shift Work by Judicious Light and Darkness Exposure." *Journal of Biological Rhythms* 17:556–567.

Bonnefond, Anne, Patricia Tassi, Joceline Roge, and Alain Muzet. 2004. "A Critical Review of Techniques Aiming at Enhancing and Sustaining Workers' Alertness During the Night." *Industrial Health* 42:1–14.

Borden, Sam. 2004. "Bombers: We're Just Plane Tired." *New York Daily News*, April 4:67.

Borst, Arno. 1993. *The Ordering of Time: From the Ancient Computus to the Modern Computer*, translated by Andrew Winnard. Chicago: University of Chicago Press.

Bourdieu, Pierre. 1963. The Attitude of the Algerian Peasant toward Time. In *Mediterranean Countrymen*, edited by Julian Pitt-Rivers, 55–72. Paris: Mouton.

———. 1977 (1972). *Outline of a Theory of Practice*, translated by Richard Nice. Cambridge: Cambridge University Press.

———. 1979 (1977). "The Disenchantment of the World." In *Algeria 1960*, translated by Richard Nice, 1–94. Cambridge: Cambridge University Press.

———. 1990 (1980). *The Logic of Practice*, translated by Richard Nice. Stanford: Stanford University Press.

Bradley, S. A. J. 2002. *Orm Gamalson's Sundial: The Lily's Blossom and the Roses' Fragrance (The 1997 Kirkdale Lecture)*. Kirkdale, Yorkshire: The Trustees of the Friends of St. Gregory's Minster.

Brennan, Martin. 1983. *The Stars and the Stones: Ancient Art and Astronomy in Ireland*. London: Thames and Hudson.

Bronson, F. H. 2004. "Are Humans Seasonally Photoperiodic?" *Journal of Biological Rhythms* 19:180–192.

Brooks, George E. 1984. "The Observance of All Souls' Day in the Guinea-Bissau Region: A Christian Holy Day, and African Harvest Festival, and African New Year's Celebration, or All of the Above(?)." *History in Africa* 11:1–34.

Bruegel, Martin. 1995. "'Time that Can be Relied Upon': The Evolution of Time Consciousness in the Mid-Hudson Valley, 1790–1860." *Journal of Social History* 28:547–564

Buddenborg, Pius. 1936. "Zur Tagesordnung in der Benediktinerregel." *Benediktinishce Monatschrift zur Pflege religiösen und geisteigen Lebens* 18:88–100.

Burman, Rickie. 1981. "Time and Socioeconomic Change on Simbo, Solomon Islands." *Man* (N.S.) 16:251–267.

Buxton, Orfeu M., Calvin W. Lee, Merielle L'Hermite-Balériaux, Fred W. Turek, and Eve van Cauter. 2003. "Exercise Elicits Phase Shifts and Acute Alterations of Melatonin that Vary with Circadian Phase." *American Journal of Physiology: Regulatory, Integrative, and Comparative Physiology* 284:R714-R724.

Byrhtferth. 1995 (ca. 1010). *Enchiridion*, translated and edited by Peter S. Baker and Michael Lapidge. Oxford: The Early English Text Society.

Cassian, John. 2000. (fourth–fifth centuries). *The Institutes*, translated by Boniface Ramsey. New York: The Newman Press.

Castells, Manuel. 2000. *The Rise of the Network Society*, second edition. Oxford: Blackwell.

Chakrabarty, Dipesh. 1997. "The Time of History and the Times of Gods." In *Politics of Culture in the Shadow of Capital*, edited by Lisa Lowe and David Lloyd, 35–60. Durham: Duke University Press.

Chatterton, P., and R. Hollands. 2002. "Theorising Urban Playscapes: Producing, Regulating and Consuming Youthful Nightlife City Spaces." *Urban Studies* 39:95–116.

Chaucer, Geoffrey. 1988a (ca. 1391). "A Treatise on the Astrolabe." In *The Riverside Chaucer*, edited by Christopher Cannon, 662–683. Oxford: Oxford University Press.

———. 1988b (ca. 1390). "The Canterbury Tales." In *The Riverside Chaucer*, edited by Christopher Cannon, 3–328. Oxford: Oxford University Press.

Cipolla, Carlo M. 1978 (1967). *Clocks and Culture, 1300–1700*. New York: Norton.

Clark, Andy. 1989. *Microcognition: Philosophy, Cognitive Science and Parallel Distributed Processing*. Cambridge, MA: MIT Press.

———. 2008. *Supersizing the Mind*. Oxford: Oxford University Press.

Clark, Andy, and David Chalmers. 1998. "The Extended Mind." *Analysis* 58:7–19.

Clark, Willene B. 1982. "The Illustrated Medieval Aviary and the Lay-Brotherhood." *Gesta* 21:63–74.

Clarke, S. R., and J. T. Norman. 2003. "Dynamic Programming in Cricket: Choosing a Night Watchman." *The Journal of the Operational Research Society* 54:838–845.

Cohn, Samuel. 2007. "After the Black Death: Labour Legislation and Attitudes Towards Labour in Late-Medieval Western Europe." *Economic History Review* 60:457–485.

Cole, Michael. 1995. "Culture and Cognitive Development: From Cross-Cultural Research to Creating Systems of Cultural Mediation." *Culture & Psychology* 1:25–54.

Comitas, Lambros. 1973. "Occupational Multiplicity in Rural Jamaica." In *Work and Family Life: West Indian Perspectives*, edited by Lambros Comitas and David Lowenthal, 156–173. Garden City, NY: Anchor.

Condon, Richard G. 1983. *Inuit Behavior and Seasonal Change in the Canadian Arctic*. Ann Arbor, MI: UMI Research Press.

Cook, Frederick, ed. 1887. *Journals of the Military Expedition of Major General John Sullivan against the Six Nations of Indians in 1779 with Records of Centennial Celebrations*. Auburn, NY: Knapp, Peck and Thomson.

Corbin, Alain. 1998. *Village Bells: Sound and Meaning in the 19th-Century French Countryside*, translated by Martin Thom. New York: Columbia University Press.

Costa, Giovanni, Torbjorn Åkerstedt, Friedhelm Nachreiner, Gederica Baltieri, José Carvalhais, Simon Folkard, Monique Frings Dresen, Charles Gadbois, Johannes Gartner, Hiltraud Grzech Sukalo, Mikko Härmä, Kandolin Irja, Samantha Sartori, and Jorge Silvério. 2004. "Flexible Working Hours, Health, and Well-Being in Europe: Some Considerations from a SALTSA Project." *Chronobiology International* 21:831–844.

Craik, Kenneth H., and Theodore R. Sarbin. 1963. "Effect of the Covert Alterations of Clock Rate upon Time Estimations and Personal Tempo." *Perceptual and Motor Skills* 16:597–610.

Crossley, F. H. 1962 *The English Abbey*. London: B. T. Batsford.

Czeisler, Charles A. 2003. "Quantifying Consequences of Chronic Sleep Restriction." *Sleep* 26:247–248.

Czeisler, Charles A., and Emery N. Brown. 1999. "Commentary: Models of the Effect of Light on the Human Circadian System: Current State of the Art." *Journal of Biological Rhythms* 14:538–543.

Czeisler, Charles A., David Dinges, James K. Walsh, Thomas Roth, and Jonathan Neibler. 2003. "Modafinil for the Treatment of Excessive Sleepiness in Chronic Shift Work Sleep Disorder." *Sleep* 26:A114.

Czeisler, Charles A., Jeanne F. Duffy, Theresa L. Shanahan, Emery N. Brown, Jude F. Mitchell, David W. Rimmer, Joseph M. Ronda, Edward J. Silva, James S. Allan, Jonathan S. Emens, Derk-Jan Dijk, and Richard E. Kronauer. 1999. "Stability, Precision, and Near-24-h Period of the Human Circadian Pacemaker." *Science* 284:2177–2181.

Czeisler, Charles A., James K. Walsh, Thomas Roth, Rod J. Hughes, Kenneth P. Wright, Lilliam Kingsbury, Sanjay Arora, Jonathan R. L. Schwartz, Gwendolyn E. Niebler, and David Dinges. 2005. "Modafinil for Excessive Sleepiness Associated with Shift-Work Sleep Disorder." *New England Journal of Medicine* 353:476–486.

Czeisler, Charles A., and Kenneth P. Wright, Jr. 1999. "Influence of Light on Circadian Rhythmicity in Humans." In *Regulation of Sleep and Circadian Rhythms*, edited by Fred W. Turek and Phyllis C. Zee, 149–180. New York: Marcel Dekker.

Dalton, G. F. 1972. "Kings Dying on Tuesday." *Folklore* 83:220–224.

Danby, Colin. 2004. "Toward a Gendered Post Keynesianism: Subjectivity and Time in a Nonmodernist Framework." *Feminist Economics* 10:55–75.

D'Andrade, Roy, and Claudia Strauss, eds. 1992. *Human Motives and Cultural Models.* Cambridge: Cambridge University Press.

Danilenko, Konstanin V., Anna Wirz-Justice, Kurt Kräuchi, Jakob M. Weber, and Michael Terman. 2000. "The Human Circadian Pacemaker Can See by the Dawn's Early Light." *Journal of Biological Rhythms* 15:437–446.

Dash, Michael. 1995. *Édouard Glissant.* Cambridge: Cambridge University Press.

de Solla Price, Derek. 1974. "Gears from the Greeks, The Antikythera Mechanism: A Calendar Computer from ca. 80 B.C." *Transactions of the American Philosophical Society* (N.S.), 64:1–70.

Deleuze, Gilles. 1994. *Difference and Repetition*, translated by Paul Patton. New York: Columbia University Press.

Derrida, Jacques. 1992. *Given Time: I. Counterfeit Money.*, translated Peggy Kamuf. Chicago: University of Chicago Press.

Dijk, Derk-Jan, and Dale M. Edgar. 1999. "Circadian and Homeostatic Control of Wakefulness and Sleep." In *Regulation of Sleep and Circadian Rhythms,* edited by Fred W. Turek and Phyllis C. Zee, 111–147. New York: Marcel Dekker.

Dinges, David F., Frances Pack, Katherine Williams, Kelly A. Gillen, John W. Powell, Geoffrey Ott, Caitlin Aptowicz, and Allan I. Pack. 1997. "Cumulative Sleepiness, Mood Disturbance, and Psychomotor Vigilance Performance Decrements during a Week of Sleep Restricted to 4–5 Hours per Night." *Sleep* 20:267–277.

Dinges, David F., N. L. Rogers, and M. D. Baynard. 2005. "Chronic Sleep Deprivation." In *Principles and Practice of Sleep Medicine,* fourth edition, edited by M. H. Kryger, T. Roth, and W. C. Dement, 67. Philadelphia: W. B. Saunders Company.

Dionysius Exiguus. 1844–64 (525). "Epistolae Duae de Ratione Paschae." In *Patrologia Latina, Volume 67*, edited by J.-P. Migne, columns 20–28. Turnhold: Typographi Brepols Editores Pontifici.

Dohrn van-Rossum, Gerhard. 1996. *The History of the Hour: Clocks and Modern Temporal Orders,* translated by Thomas Dunlap. Chicago: University of Chicago Press.

Duffy, Eamon. 2006. *Marking the Hours: English People and Their Prayers 1240–1570*. New Haven: Yale University Press.

Dumont, Marie, Dalila Benhaberou-Brun, and Jean Paquet. 2001. "Profile of 24-h Light Exposure and Circadian Phase of Melatonin Secretion in Night Workers." *Journal of Biological Rhythms* 16:502–511.

Dunlap, Jay C., Jennifer J. Loros, and Patricia J. Decoursey. 2004. *Chronobiology: Biological Timekeeping*. Sunderland, MA: Sinauer.

Dunn, Frances M. 1998. "Tampering with the Calendar." *Zeitschrift für Papyrologie und Epigraphik* 123:213–231.

———. 1999. "The Council's Solar Calendar." *The American Journal of Philology* 120:369–380.

Durkheim, Émile. 2001 (1912). *The Elementary Forms of Religious Life*, translated by Carol Cosman. Oxford: Oxford University Press.

Eastman, Charmane, Erin Hoese, Shawn Youngstedt, and Liwen Liu. 1995. "Phase-Shifting Human Circadian Rhythms With Exercise during the Night Shift." *Physiology and Behavior* 58:1287–1291.

Eastman, Charmane, and Stacia Martin. 1999. "How to Use Light and Dark to Produce Circadian Adaptation to Night Shift Work." *Annals of Medicine* 31:87–98.

Einstein, Alfred. 1992 (1926). "Space-Time." In *The Treasury of the Encyclopedaeia Britannica*, edited by Clifton Fadiman, 371–383. New York: Viking.

Elias, Norbert. 1992. *Time: An Essay*. Oxford: Blackwell.

Epstein, Steven A. 1988. "Business Cycles and the Sense of Time in Medieval Genoa." *The Business History Review* 62:238–260.

Essen, Louis. 2000. "The Atomic Clock." In *Time for Reflection*. Last modified December 4. http://www.btinternet.com/~time.lord/.

Evans-Pritchard, E. E. 1940. *The Nuer: A Description of the Modes of Livelihood and Political Institutions of a Nilotic People*. Oxford: Oxford University Press.

Fabian, Johannes. 2002 (1983). *Time and the Other: How Anthropology Makes Its Object*. New York: Columbia University Press.

Feeney, Denis. 2007. *Caesar's Calendar: Ancient Time and the Beginning's of History*. Berkeley: University of California Press.

Fetterman, J. Gregor, and Peter R. Killeen. 1990. "A Componential Analysis of Pacemaker-Counter Timing Systems." *Journal of Experimental Psychology: Human Perception and Behavior* 16:766–780.

Finkleman, David, Steve Allen, John H. Seago, Rob Seaman, and P. Kenneth Seidelmann. 2011. "The Future of the Leap Second." *American Scientist* 99:312–319.

Fish, Stanley. 1980. *Is There a Text in This Class: The Authority of Interpretive Communities*. Cambridge, MA: Harvard University Press.

Flamsteed, John. 1997 (1700). *The Correspondence of John Flamsteed, First Astronomer Royal, Volume 2*, edited by Eric G. Forbes, Lesley Murdin, and Frances Willmoth. Bristol: Institute of Physics Publishing.

Folkard, Simon. 1997. "Black Times: Temporal Determinants of Transport Safety." *Accident Analysis and Prevention* 4:417–430.

Folkard, Simon, Torbjorn Åkerstedt, Ian MacDonald, Philip Tucker, and Michael Spencer. 1999. "Beyond the Three-Process Model of Alertness: Estimating Phase, Time on Shift, and Successive Night Effects." *Journal of Biological Rhythms* 14:577–587.

Folkard, Simon, and Jane Barton. 1993. "Does the 'Forbidden Zone' for Sleep Onset Influence Morning Shift Sleep Duration?" *Ergonomics* 36:85–91.

Foster, Russell G., and Leon Kreitzmann. 2004. *Rhythms of Life*. New Haven: Yale University Press.

Foster, Russell G., and Till Roenneberg. 2008. "Human Responses to the Geophysical Daily, Annual and Lunar Cycles." *Current Biology* 18:R784–R794.

Foucault, Michel. 1977. *Discipline and Punish: The Birth of the Prison*, translated by Alan Sheridan. London: Allen Lane.

Frake, Charles. 1985. "Cognitive Maps of Time and Tide among Medieval Seafarers." *Man* (N.S.) 20:254–270.

Freud, Sigmund. 1976 (1919). "The 'Uncanny,'" translated by James Strachey. *New Literary History* 7:619–645.

Friedman, Thomas. 2005. *The World is Flat: A Brief History of the Twenty-First Century*. New York: Farrar, Straus and Giroux.

Galison, Peter Louis. 2003. *Einstein's Clocks and Poincaré's Maps: Empires of Time*. New York: W. W. Norton.

Gallicchio, Lisa, and Bindu Kalesan. 2008. "Sleep Duration and Mortality: A Systematic Review and Meta-Analysis." *Journal of Sleep Research* 18:148–158.

Geertz, Clifford. 1973. *The Interpretation of Cultures*. New York: Basic Books.

Gell, Alfred. 1975. *Metamorphosis of the Cassowaries: Umeda Society, Language, and Ritual*. London: Althone.

———. 1992. *The Anthropology of Time*. Oxford: Berg.

Gellius, Aulus. 1927 (second century). *The Attic Nights, Volume 1*, translated by John C. Rolfe. London: Loeb Classical Library.

Giebel, Ole, Daniela Janßen, Carsten Schomann, and Friedhelm Nachreiner. 2004. "a New Approach for Evaluating Flexible Working Hours." *Chronobiology International* 21:1015–1024.

Gilson, Etienne. 1955. *History of Christian Philosophy in the Middle Ages*. New York: Random House.

Gladwin, Thomas. 1970. *East is a Big Bird: Navigation and Logic on Puluwat Atoll*. Cambridge: Harvard University Press.

Glasser, Richard. 1972 . *Time in French Life and Thought*, translated by C. G. Pearson. Totowa, NJ: Rowman and Littlefield.

Glennie, Paul, and Nigel Thrift. 2002. "The Spaces of Clock Times." In *The Social in Question: New Bearings in History and the Social Sciences*, edited by Patrick Joyce, 151–174. Routledge: London.

———. 2005. "Revolutions in the Times: Clocks and the Temporal Structures of Everyday Life." In *Geography and Revolution*, edited by D. Livingstone and C. Withers, 160–198. Chicago: University of Chicago Press.

———. 2009. *Shaping the Day: A History of Timekeeping in England and Wales 1300–1800*. Oxford: Oxford University Press.

Glissant, Édouard. 1956. *Soleil de la conscience*. Paris: Seuil.

———. 1989. *Caribbean Discourse*, translated by J. Michael Dash. Charlottesville: University of Virginia Press.

Good, Anthony. 2000. "Congealing Divinity: Time, Worship and Kinship in South Indian Hinduism." *Journal of the Royal Anthropological Institute* 6:273–292.

Goodman, Jane. 2003. "The Proverbial Bourdieu: Habitus and the Politics of Representation in Ethnography of Kabylia." *American Anthropologist* 105:782–293.

Goody, Jack, and Ian Watt. 1963. "The Consequences of Literacy." *Comparative Studies in Society and History* 5:304–345.

Greenhouse, Carol. 1996. *A Moment's Notice*. Ithaca: Cornell University Press.

Gregory the XIII, Pope. 2002 (1582) *Inter Gravissimus*. Last modified March 20, 2002. http://www.bluewaterarts.com/calendar/NewInterGravissimas.htm.

Gregory the Great, Pope. 1850 (sixth century). *Morals on the Book of Job, Volume 3, Part 2, Books 30–35*. Oxford: John Henry Parker.

Gupta, Akhil. 1994. "The Reincarnation of Souls and the Rebirth of Commodities: Representations of Time in 'East' and 'West.'" In *Remapping Memory: The Politics of TimeSpace*, edited by Johnathan Boyarin, 161–183. Minneapolis: University of Minnesota Press.

Gurevich, A. J. 1976. "Time as a Problem of Cultural History." In *Cultures and Time*, edited by Louis Gardet, 229–245. Paris: The Unesco Press.

Hallowell, A. Irving. 1955. *Culture and Experience*. Philadelphia: University of Pennsylvania Press.

Hamermesh, Daniel S. 1999. "The Timing of Work over Time." *The Economic Journal* 109:37–66.

Hannah, Robert. 2001. "From Orality to Literacy? The Case of the Parapegma." In *Speaking Volumes: Orality and Literacy in the Greek and Roman World*, edited by Janet Watson, 139–159. Leiden Brill.

———. 2008. "Timekeeping." In *Oxford Handbook of Engineering and Technology in the Classical World*, edited by John Peter Oleson, 740–758. Oxford: Oxford University Press.

———. 2009. *Time in Antiquity*. New York: Routledge.

Harris, Mark. 1998. "The Rhythm of Life on the Amazon Floodplain: Seasonality and Sociality in a Riverine Village." *Journal of the Royal Anthropological Institute* 4:65–82.

Harris, Wilson. 1999. *Selected Essays of Wilson Harris*, edited by Andrew Bundy. New York: Routledge.

Harvey, David. 1989. *The Condition of Postmodernity: An Enquiry into the Origins of Culture Change*. Oxford: Blackwell.

———. 1990. "Between Space and Time: Reflections on the Geographical Imagination." *Annals of the Association of American Geographers* 80:418–434.

———. 1993. "From Space to Place and Back Again: Reflections on the Condition of Postmodernity." In *Mapping the Futures: Local Cultures, Global Change*, edited by Jon Bird, Barry Curtis, Tim Putnam, George Robertson, and Lisa Tickner, 3–29. London: Routledge.

Hassan, Robert, and Ronald E. Purser. 2007. Introduction to *24/7: Time and Temporality in the Network Society*, 1–36. Palo Alto, CA.: Stanford University Press.

Hesiod. 1998 (ca. eighth century BC). *Theogony and Works and Days*, translated by M. L. West. Oxford: Oxford University Press.

Holmes, Urban Tigner. 1964. *Daily Living in the Twelfth Century: Based on the Observations of Alexander Neckam in London and Paris*. Madison: University of Wisconsin Press.

Hongaldarom, Soraj. 2002. "The Web of Time and the Dilemma of Globalization." *The Information Society* 18:241–249.

Hosack, William. 1986 (1879). Excerpt from "The Isle of Streams, or, the Jamaican Hermit." In *The Penguin Book of Caribbean Verse in English*, edited by Paula Burnett, 117–120. New York: Penguin.

Hoskins, Janet. 1997. *The Play of Time: Kodi Perspectives on Calendars, History, and Exchange*. Berkeley: University of California Press.

Howse, Derek. 1980. *Greenwich Time and the Discovery of Longitude*. Oxford: Oxford University Press.

Hubert, Henri. 1999 (1909). *Essay on Time: A Brief Study of the Representation of Time in Religion and Magic*, translated by Robert Parkin and Jacqueline Redding. Oxford: Durkheim Press.

Hugo de Folieto. 1844–64 (twelfth century). "De Bestiis et aliis Rebus Libri Quatuor." In *Patrologia Latina, Volume 177*, edited by J.-P. Migne, columns 9–164. Turnhold: Typographi Brepols Editores Pontifici.

Hutchins, Edwin. 1995. *Cognition in the Wild*. Cambridge: MIT Press.

———. 1999. "Cognitive Artifacts." In *MIT Encyclopedia of the Cognitive Sciences*, edited by Robert A. Wilson and Frank C. Keil, 126–127. Cambridge, MA: MIT Press.

———. 2008. "The Role of Cultural Practices in the Emergence of Modern Human Intelligence." *Philosophical Transactions of the Royal Society, B* 363:2011–2019.

Hutton, Ronald. 1996. *The Stations of the Sun: A History of the Ritual Year in Britain*. Oxford: Oxford University Press.

Institute of Medicine. 2006. *Sleep Disorders and Sleep Deprivation: An Unmet Public Health Problem*. Washington, DC: The National Academies Press.

Isidore of Seville. 2005 (seventh century). *Etymologies*, translated by Priscilla Throop. Charlotte, VT: MedievalMS.

ITU-R. 2011. Status of Coordinate Universal Time (UTC) study in ITU-R. Accessed November 15, 2011. http://www.itu.int/oth/R0A08000009/en.

James, C. L. R. 1993 (1963). *Beyond a Boundary*. Durham, NC: Duke University Press.

Jameson, Fredric. 1991. *Postmodernism or, the Cultural Logic of Late Capitalism*. Durham: Duke University Press.

Janssen, Daniela, and Friedhelm Nachreiner. 2004. "Health and Psychosocial Effects of Flexible Working Hours." *Revista de Saúde Pública* 38 (supplement):11–18.

Jones, Charles. 1943. "Development of the Latin Ecclesiastical Calendar." In *Bedae Opera de Temporibus*, 3–121. Cambridge, MA: The Mediaeval Academy of America.

Justeson, John S. 1989. "Ancient Maya Ethnoastronomy: Between Two Worlds." In *World Archaeoastronomy: Selected Papers from the 2nd Oxford International Conference on Archaeoastronomy*, edited by Anthony Aveni, 76–129. Cambridge: Cambridge University Press.

Kanigel, Robert. 1997. *One Best Way: Frederic Winslow Taylor and the Enigma of Efficiency*. New York: Viking.

Kelly, Tamsin Lisa, David F. Neri, Jeffrey T. Grill, David Ryman, Phillip D. Hunt, Derk- Jan Dijk, Theresa L. Shanahan, and Charles A. Czeisler. 1999. "Nonentrained Circadian Rhythms of Melatonin in Submariners Scheduled to an 18-hour Day." *Journal of Biological Rhythms* 14:190–196.

Kelsey, Harry. 1983. "The Gregorian Calendar in New Spain: A Problem in Sixteenth-Century Chronology." *New Mexico Historical Review* 58:239–252.

Kirsch, David, and Paul Maglio. 1994. "On Distinguishing Epistemic from Pragmatic Action." *Cognitive Science* 18:513–549.

Klass, Morton. 1979. *From Field to Factory: Community Structure and Industrialization in West Bengal*. Philadelphia: Institute for the Study of Human Issues.

Klein, Karl E., and Hans-Martin Wegmann. 1974. "The Resynchronization of Human Circadian Rhythms after Transmeridian Flights as a Result of Flight Direction and Mode of Activity." In *Chronobiology*, edited by Lawrence E. Scheving, Franz Halberg, and John E. Pauly, 564–575. Tokyo: Igaku Shoin Ltd.

Knowles, Dom David. 1951. Introduction to *The Monastic Constitutions of Lanfranc*. London: Thomas Nelson and Sons.

———. 1966. *The Monastic Order in England: A History of its Development from the Times of St. Dunstan to the Fourth Lateran Council, 940–1216,* second edition. Cambridge: Cambridge University Press.

Knutsson, Anders. 2003. "Health Disorders of Shift Workers." *Occupational Medicine* 53:103–108.

Koller, M., M. Härma, J. T. Laitinen, M. Kundi, B. Piegler, and M. Haider. 1994. "Different Patterns of Light Exposure in Relation to Melatonin and Cortisol Rhythms and Sleep of Night Workers." *Journal of Pineal Research* 16:127–135.

Kramsch, Claire. 2004. "Language, Thought and Culture." In *The Handbook of Applied Linguistics*, edited by Allan Davies and Catherine Elder, 235–261. Oxford: Blackwell.

LaClau, Ernesto. 1990. *New Reflections on the Revolution of Our Time*, translated by Jon Barnes. New York: Verso.

Landes, David. 1983. *Revolution in Time: Clocks and the Making of the Modern World*. Cambridge, MA: Belknap.

Lanfranc. 1951 (eleventh century). *The Monastic Constitutions*, translated by David Knowles. London: Thomas Nelson.

Latour, Bruno. 1993. *We Have Never Been Modern*, translated by Catherine Porter. Cambridge, MA: Harvard University Press.

———. 1997. "Trains of Thought: Piaget, Formalism, and the Fifth Dimension." *Common Knowledge* 6:170–191.

Latour, Bruno, and Steve Woolgar. 1986. *Laboratory Life: The Construction of Scientific Facts*. Princeton: Princeton University Press.

Lave, Jean. 1988. *Cognition in Practice: Mind, Mathematics and Culture in Everyday Life*. Cambridge: Cambridge University Press.

Lave, Jean, and Etienne Wenger. 1991. *Situated Learning: Legitimate Peripheral Participation*. Cambridge: Cambridge University Press.

Le Goff, Jacques. 1980. *Time, Work, and Culture in the Middle Ages*, translated by Arthur Goldhammer. Chicago: University of Chicago Press.

———. 1988. *Medieval Civilization, 400–1500*. Translated by Julia Barrow. Oxford: Blackwell.

Leach, Edmund. 1950. "Primitive Calendars." *Oceania* 20:245–262.

———. 1961. *Rethinking Anthropology*. London: The Athlone Press.

LeFebvre, Henri. 2004 (1992). *Rhythmanalysis: Space, Time and Everyday Life*, translated by Stuart Elden and Gerald Moore. London: Continuum.

Lehoux, Daryn. 2004. "Observation and Prediction in Ancient Astrology." *Studies in the History and Philosophy of Science* 35:227–246.

———. 2007. *Astronomy, Weather and Calendars in the Ancient World: Parapegmata and Related Texts in Classical and Near Eastern Societies*. Cambridge: Cambridge University Press.

Leibniz, Gottfried Wilhelm. 1997 (1700). "Leibniz to the Royal Society 30 January 1699/1700." In *The Correspondence of John Flamsteed, First Astronomer Royal, Volume II*, edited by Eric G. Forbes, Lesley Murdin, and Frances Willmoth, 805–810. Bristol: Institute of Physics Publishing.

Leont'ev, A. N. 1979 (1972). "The Problem of Activity in Psychology." In *The Concept of Activity in Soviet Psychology*, translated and edited by James V. Wertsch, 37–71. Armonk, NY: M. E. Sharpe.

———. 1997. "On Vygotsky's Creative Development." In *The Collected Works of L. S. Vygotsky, Volume 3: Problems of the Theory and History of Psychology*, edited by Robert W. Reiber and Jeffrey Wollock, 9–32. New York: Plenum.

Leschiutta, S. 2005. "The Definition of the Atomic Second." *Metrologia* 42:S10–19.

Linnaeus, Carolus. 2003 (1751). *Philosophia Botanica*, translated by Stephen Freer. Oxford: Oxford University Press.

Loewe, Michael. 1999. "Cyclical and Linear Concepts of Time in China." In *Time*, edited by Kristen Lippencott, 76–77. London: Merrell Holberton.

Lucy, John A. 1992. *Language Diversity and Thought: A Reformulation of the Linguistic Relativity Hypothesis*. Cambridge: Cambridge University Press.

———. 1997. "Linguistic Relativity." *Annual Review of Anthropology* 26:291–312.

Lynds, Peter. 2003. "Time and Classical and Quantum Mechanics: Indeterminacy vs. Discontinuity." *Foundations of Physics Letters* 16:343–355.

Macrobius. 2011 (fifth century). *Saturnalia*, translated and edited by Robert A. Kaster. Cambridge, MA: Loeb Classical Library.

Magee, C. A., D. C. Iverson, X. F. Huang, and P. Caputi. 2008. "A Link between Chronic Sleep Restriction and Obesity: Methodological Considerations." *Public Health* 22:1373–1381.

Malinowski, Bronislaw. 1927. "Lunar and Seasonal Calendar in the Trobriands." *The Journal of the Royal Anthropological Institute of Great Britain and Ireland* 57:203–215.

———. 1935. *Coral Gardens and Their Magic: A Study of the Methods of Tilling the Soil and of Agricultural Rites in the Trobriand Islands*. London: Allen and Unwin.

———. 1961 (1922). *Argonauts of the Western Pacific*. New York: Dutton.

Margolis, Eric, and Stephen Laurence, eds. 2007. *Creations of the Mind: Theories of Artifacts and Their Representations*. Oxford: Oxford University Press.

Markowitz, W., R. G. Hall, L. Essen, and J. V. L. Perry. 1958. "Frequency of Cesium in Terms of Ephemeris Time." *Physical Review Letters* 1:105–107.

Marshack, Alexander. 1972. *The Roots of Civilization: The Cognitive Beginnings of Man's First Art, Symbol and Notation*. New York: McGraw Hill.

Marx, Karl. 1970 (1859). *A Contribution to the Critique of Political Economy*, translated by S. W. Ryazanskaya. Moscow: Progress Publishers.

———. 1977 (1867). *Capital*, translated by Ben Fowkes. New York: Vintage.

Massey, Doreen. 1992. "Politics and Space/Time." *New Left Review* 19:65–84.

McCarthy, Dennis D., and P. Kenneth Seidelmann. 2009. *Time—From Earth Rotation to Atomic Physics.* Weinheim: Wiley-VCH.

McCluskey, Stephen C. 1998. *Astronomies and Culture in Early Modern Europe.* Cambridge: Cambridge University Press.

McEachron, Donald L., and Jonathan Schull. 1993. "Hormones, Rhythms, and the Blues." In *Hormonally Induced Changes in Mind and Brain*, edited by Jay Schulkin, 287–355. San Diego: Academic.

Meerlo, Peter, Ralph E. Mistlberger, Barry L. Jacobs, Craig H. Heller, and Dennis McGinty. 2008. "New Neurons in the Adult Brain: The Role of Sleep and Consequences of Sleep Loss." *Sleep Medicine Reviews* 13:187–194.

Meijer, Anne Marie. 2008. "Chronic Sleep Reduction, Functioning at School and School Achievement in Preadolescents." *Journal of Sleep Research* 17:395–405.

Meijer, Johanna H., and William J. Schwartz. 2003. "In Search of the Pathways for Light-Induced Pacemaker Resetting in the Suprachiasmatic Nucleus." *Journal of Biological Rhythms* 18:235–249.

Miller, Daniel. 1994. *Modernity—an Ethnographic Approach: Dualism and Mass Consumption in Trinidad.* Oxford: Berg.

———. 1997. *Capitalism: An Ethnographic Approach.* Oxford: Berg.

———. 2010. *Stuff.* Malden, MA: Polity Press.

Mintz, Sidney. 1985. *Sweetness and Power: The Place of Sugar in Modern History.* New York: Penguin.

———. 1993. "Enduring Substances, Trying Theories: The Caribbean Region as Oikoumenê." *Journal of the Royal Anthropological Institute* 2:289–311.

Mishnah Berakoth. 1948 (ca. 200). *The Bablylonian Talmud—Seder Sera'im: Berakoth*, translated by Maurice Simon. London: Soncino Press.

Mistlberger, Ralph E., and Debra J. Skene. 2005. "Nonphotic Entrainment in Humans?" *Journal of Biological Rhythms* 20:339–352.

Mondragón, Carlos. 2004. "Of Winds, Worms and Mana: The Traditional Calendar of the Torres Islands, Vanuatu." *Oceania* 74:289–308.

Monk, Timothy H. 2000. "What Can the Chronobiologist Do to Help the Shift Worker?" *Journal of Biological Rhythms* 15:86–94.

Monk, Timothy H., Daniel J. Buysse, Bart D. Billy, Kathy S. Kennedy, and Linda M. Willrich. 1998. "Sleep and Circadian Rhythms in Four Orbiting Astronauts." *Journal of Biological Rhythms*, 13:188–201.

Monteleone, Palmiero, Antonio Fuschino, Giovanni Nolfe, and Mario Maj. 1992. "Temporal Relationship between Melatonin and Cortisol Responses to Nighttime Physical Stress in Humans." *Psychoneuroendocrinology* 17:81–86.

Mooney, Linne R. 1993. "The Cock and the Clock: Telling Time in Chaucer's Day." *Studies in the Age of Chaucer* 15:91–109.

Moore-Ede, Martin C. 1993. *The 24-Hour Society: Understanding Human Limits in a World that Never Stops*. New York: Addison-Wesley.

Moore-Ede, Martin C., Frank M. Sulzman, and Charles A. Fuller. 1982. *The Clocks that Time Us: Physiology of the Circadian Timing System*. Harvard: Harvard University Press.

Mrosovsky, N. 1996. "Locomotor Activity and Non-photic Influences on Circadian Clocks." *Biological Reviews of the Cambridge Philosophical Society* 71:343–372.

Mrosovsky, N., S. G. Reebs, G. I. Honrado, and P. A. Salmon. 1989. "Behavioral Entrainment of Circadian Rhythms." *Experientia* 45:696–702.

Mumford, Lewis. 1963 (1934). *Technics and Civilization*. New York: Harcourt, Brace and World.

Munn, Nancy. 1986. *The Fame of Gawa: A Symbolic Study of Value Transformation in a Massim (Papua New Guinea) Society*. Cambridge: Cambridge University Press.

———. 1992. "The Cultural Anthropology of Time: A Critical Essay." *Annual Reviews of Anthropology* 21:93–123.

Murray, Louise Welles. 1975 (1930). *Notes from Collections of Tioga Point Museum on the Sullivan Expedition of 1779 and its Centennial Celebration of 1879*. Athens, PA: Tioga Point Museum.

Neckam, Alexander. 1863 (ca. 1180). *De Naturis Rerum et De Laudibus Divinae Sapientiae*. London: Longman, Green, Longman, Roberts, and Green.

Needham, Joseph. 1959. *Science and Civilisation in China, Volume 3: Mathematics and the Sciences of the Heavens and the Earth*. Cambridge: Cambridge University Press.

Newcomb, Simon. 1895. *Astronomical Papers Prepared for the Use of the American Ephemeris and Nautical Almanac, Volume VI, Part I: Tables of the Sun*. Washington, DC: US Government Printing Office.

Newton, Isaac. 1934 (1687). *Sir Isaac Newton's Mathematical Principles of Natural Philosophy and his System of the World*. Berkeley: University of California Press.

———. 1998 (1733). *The Prophecies of Daniel and the Apocalypse.* Hyderabad, India: Printland Publishers.

New York Times. 1928. "Watches Tested for Their Whims." *New York Times*, January 8, 124.

Nietschze, Friedrich. 1967 (1872). *The Birth of Tragedy, and The Case of Wagner.* New York: Vintage.

Noë, Alva. 2009. *Out of Our Heads: Why You Are Not Your Brain, and Other Lessons from the Biology of Consciousness.* New York: Hill and Wang.

North, J. D. 1983. "The Western Calendar—'Intolerabilis, Horribilis, et Derisibilis': Four Centuries of Discontent." In *Gregorian Reform of the Calendar*, edited by G. V. Coyne, M. A. Hoskin, and O. Pedersen, 75–113. Vatican: Pontificia Academia Scientiarum.

———. 1988. *Chaucer's Universe.* Oxford: Clarendon Press.

O'Connor, Anahad. 2004. "Wakefulness Finds a Powerful Ally." *New York Times*, June 29, F1.

O'Malley, Michael. 1990. *Keeping Watch: A History of American Time.* New York: Penguin.

O'Neil, W. M. 1975. *Time and the Calendars.* Syndey: Sydney University Press.

Ong, Walter J. 1982. *Orality and Literacy: The Technologizing of the Word.* New York: Methuen.

Ortner, Sherry. 2006. *Anthropology and Social Theory: Culture, Power, and the Acting Subject.* Durham, NC: Duke University Press.

Otto, Rudolf. 1950 (1917). *The Idea of the Holy: An Inquiry into the Non-Rational Factor in the Idea of the Divine and Its Relation to the Rational*, translated by John W. Harvey. Oxford: Oxford University Press.

Ovid. 1995 (ca. 8–14). *Ovid's Fasti: Roman Holidays*, translated by Betty Rose Nagle. Bloomington, IN: Indiana University Press.

Palmer, John D. 2002. *The Living Clock: The Orchestrator of Biological Rhythms.* Oxford University Press.

Peirce, Charles Sanders. 1955. *Philosophical Writings of Peirce*, edited by Justus Buchler. New York: Dover.

Peters, Jacqueline D., Sarah N. Biggs, Katie M. M. Bauer, Kurt Lushington, Declan Kennedy, James Martin, and Jillian Dorrian. 2009. "The Sensitivity of a PDA-based Psychomotor Vigilance Task to Sleep Restriction in 10-year-old Girls." *Journal of Sleep Research* 18:173–177.

Pickering, Kathleen. 2004. "Decolonizing Time Regimes: Lakota Conceptions of Work, Economy, and Society." *American Anthropologist* 106:85–97.

Pittendrigh, Colin S. 1993. "Temporal Organization: Reflections of a Darwinian Clock-Watcher." *Annual Review of Physiology* 55:17–54.

Pliny the Elder. 2005 (ca. 77–79). *On the Human Animal, Book 7: Natural History*, translated by Mary Beagon. Oxford: Oxford University Press.

Pliny the Younger. 2006 (ca. 100). "To His Friend Arrianus." In *Complete Letters.* Translated by P. G. Walsh, 38–42. Oxford: Oxford University Press.

Plutarch. 2011 (first century). *Caesar*, translated by Christopher Pelling. Oxford: Oxford University Press.

Poincaré, Henri. 1913 (1905). "Time and Its Measure." In *The Foundations of Science*, translated by George Bruce Halsted, 223–234. New York: Science Press.

Ponko, Vincent, Jr. 1968. " The Privy Council and the Spirit of Elizabethan Economic Management, 1558–1603." *Transactions of the American Philosophical Society, New Series* 58:1–63.

Poole, Robert. 1995. "'Give Us Our Eleven Days!': Calendar Reform in Eighteenth-century England." *Past and Present* 149:95–139.

———. 1998. *Time's Alteration: Calendar Reform in Early Modern England*, London: UCL Press.

Pope, Alexander. 1970 (1711). *An Essay on Criticism.* Menston, England: The Scolar Press.

Postill, John. 2002. "Clock and Calendar Time: A Missing Anthropological Problem." *Time and Society* 11:251–270.

Postone, Moishe. 2003. *Time, Labor, and Social Domination: A Reinterpretation of Marx's Critical Theory.* Cambridge: Cambridge University Press.

Prerau, David. 2005. *Seize the Daylight: The Curious and Contentious Story of Daylight Saving Time.* New York: Thunder's Mouth Press.

Presser, Harriet B. 1999. "Toward a 24-Hour Economy." *Science* 284:1778–1779.

Pritchett, W. K., and O. Neugebauer. 1947. *The Calendars of Athens.* Cambridge, MA: Harvard University Press.

Puleston, Dennis E. 1979. "An Epistemological Pathology and the Collapse, or Why the Maya Kept the Short Count." In *Maya Archaeology and Ethnohistory*, edited by Norman Hammond and Gordon R. Willey, 63–71. Austin: University of Texas Press.

Putnam, Bertha Haven. 1908. *The Enforcement of the Statute of Labourers During the First Decade after the Black Death, 1349–1359.* New York: Columbia University

Putter, Ad. 2001. "In Search of Lost Time: Missing Days in *Sir Cleges* and *Sir Gawain and the Green Knight.*" In *Time in the Medieval*

World, edited by Chris Humphrey and W. M. Omrod, 119–136. York: York Medieval Press.

Quinn, Naomi, and Dorothy Holland. 1987. "Culture and Cognition." In *Cultural Models in Language and Thought*, edited by Dorothy Holland and Naomi Quinn, 3–40. Cambridge: Cambridge University Press.

Rajaratnam, Shantha M. W., and Josephine Arendt. 2001. "Health in a 24-h Society." *The Lancet* 358:999–105.

Ratner, Carl. 2001. "Activity Theory and Cultural Psychology." In *The Psychology of Cultural Experience*, edited by Carmella C. Moore and Holly F. Mathews, 68–80. Cambridge: Cambridge University Press.

Recht, Lawrence, Robert A. Lew, and William J. Schwartz. 1995. "Baseball Teams Beaten by Jet Lag." *Nature* 377:583.

Regularis Concordia. 1953. *Regularis Concordia Angliciae Nationis Monachorum Sanctimonialiumque*, translated by Dom Thomas Symons. London: Thomas Nelson and Sons.

Reiche, Harold A. T. 1989. "Fail-Safe Stellar Dating: Forgotten Phases." *Transactions of the American Philological Association* 119:37–53.

Revell, Victoria L., and Charmane I. Eastman. 2005. "How to Trick Mother Nature into Letting You Fly Around or Stay Up All Night." *Journal of Biological Rhythms* 20:353–365.

Rice, Prudence. 2004. *Maya Political Science: Time, Astronomy, and the Cosmos*. Austin: University of Texas Press.

———. 2007. *Maya Calendar Origins: Monuments, Mythistory, and Materialization of Time*. Austin: University of Texas Press.

Robertson, Roland. 1995. "Glocalization: Time-space and Homogeneity-heterogeneity." In *Global Modernities*, edited by Mike Featherstone, Scott Lash, and Roland Robertson, 25–44. London: Sage.

Rolle, Richard (incorrect attribution). 1910 (fourteenth century). "Our Daily Work (A Mirror of Discipline)." In *The Form of Perfect Living and Other Prose Treatises*, edited by Geraldine E. Hodgson, 83–168. London: Thomas Baker.

Rosen, David M., and Victoria P. Rosen. 2000. "New Myths and Meanings in Jewish New Moon Rituals." *Ethnology* 39:263–277.

Rothwell, W. 1959. "The Hours of the Day in Medieval French." *French Studies* 13:240–251.

Rotter, George S. 1969. "Clock-Speed as an Independent Variable in Psychological Research." *The Journal of General Psychology* 81:45–52.

Rubin, David C. 1995. *Memory in Oral Traditions: The Cognitive Psychology of Epic, Ballads, and Counting-Out Rhymes*. Oxford: Oxford University Press.

Rüger, Melanie, and Frank A. J. L. Scheer. 2009. "Effects of Circadian Disruption on the Cardiometabolic System." *Reviews in Endocrine and Metabolic Disorders* 10 (4, special issue):245–260.

Rüpke, Jörg. 2011 (1995). *The Roman Calendar from Numa to Constantine: Time, History and the Fasti*. Chichester, West Sussex, UK: Wiley-Blackwell.

Salzman, L. F. 1952. *Building in England Down to 1540: A Documentary History*. Oxford: Clarendon Press.

Sapir, Edward. 1949 (1929). "The Status of Linguistics as a Science." In *Selected Writings in Language, Culture and Personality*, edited by David G. Mandelbaum, 160–166. Berkeley: University of California Press.

———. 1964 (1931). "Conceptual Categories in Primitive Languages." In *Language in Culture and Society: A Reader in Linguistics and Anthropology*, edited by D. H. Hymes, 128. New York: Harper and Row.

Sapolsky, Robert M. 1998. *Why Zebras Don't Get Ulcers*, second edition. New York: W. H. Freeman.

Sassen, Saskia. 1998. *Globalization and Its Discontents*. New York: W. W. Norton.

———. 2000. "Spatialities and Temporalities of the Global: Elements for a Theorization." *Public Culture* 12:215–232.

Sauter, Michael J. 2007. "Clockwatchers and Stargazers: Time Discipline in Early Modern Berlin." *American Historical Review* 112:685–709.

Schwartz, Theodore. 1978. "Where is the Culture?" In *The Making of Psychological Anthropology*, edited by George Spindler, 419–441. Berkeley: University of California Press.

Schwartz, William J., Horacio O. de la Iglesia, Piotr Zlomanczuk, and Helena Illenová. 2001. "Encoding *Le Quattro Stagioni* within the Mammalian Brain: Photoperiodic Orchestration through the Suprachiasmatic Nucleus." *Journal of Biological Rhythms* 16:302–311.

Searle, John. 2007. "Social Ontology and the Philosophy of Society." In *Creations of the Mind: Theories of Artifacts and their Representation*, edited by Eric Margolis and Stephen Laurence, 3–17. Oxford: Oxford University Press.

Sennett, Richard. 1998. *The Corrosion of Character: The Personal Consequences of Work in the New Capitalism*. New York: W. W. Norton.

Shackle, G. L. S. 1967. *Time in Economics*. Amsterdam: North-Holland Publishing.

Shanahan, Theresa L., Jamie M. Zeitzer, and Charles A. Czeisler. 1997. "Resetting the Melatonin Rhythm with Light in Humans." *Journal of Biological Rhythms* 12:556–567.

Shaw, Pamela-Jane. 2003. *Discrepancies in Olympiad Dating and Chronological Problems of Archaic Peloponnesian History*. Stuttgart: Franz Steiner Verlag.

Shepherd, C. Y. 1935. "Agricultural Labour in Trinidad." *Tropical Agriculture* 12:3–9, 43–47, 56–64, 84–88, 126–131, 153–157, 187–192.

Shore, Bradd. 1996. *Culture in Mind: Cognition, Culture and the Problem of Meaning*. Oxford: Oxford University Press.

Siffre, Michel. 1975. "Six Months Alone in a Cave." *National Geographic* 147:426–435.

Smith, G. Elliott, Bronislaw Malinowski, Herbert J. Spinden, and Alexander Goldenweiser. 1927. *Culture—The Diffusion Controversy*. New York: W. W. Norton.

Smith, Mark. 1997. *Mastered by the Clock: Time, Slavery and Freedom in the American South*. Chapel Hill: University of North Carolina Press.

———. 1998. "Culture, Commerce, and Calendar Reform in Colonial America." *The William and Mary Quarterly* (third series) 55:557–584.

———. 2001. "Remembering Mary, Shaping Revolt: Reconsidering the Stono Rebellion." *The Journal of Southern History* 67:513–534.

Smith, T. C. 1986. "Peasant Time and Factory Time in Japan." *Past and Present* 111:165–197.

Sobel, Dava. 1995. *Longitude: The True Story of a Lone Genius Who Solved the Greatest Scientific Problem of his Time*. New York: Penguin.

Sorokin, Pitirim, A., and Robert K. Merton. 1937. "Social Time: A Methodological and Functional Analysis." *The American Journal of Sociology* 42:615–629.

Stenuit, Patricia, and Myriam Kerkhofs. 2008. "Effects of Sleep Restriction on Cognition in Women." *Biological Psychology* 77:81–88.

Stephen, Alexander M. 1969 (1936). *Hopi Journal*, edited by Elsie Clews Parsons. New York: AMS Press.

Stern, Pamela. 2003. "Upside-Down and Backwards: Time Discipline in a Canadian Inuit Town." *Anthropologica* 45:147–161.

Stern, Sacha. 2003. *Time and Process in Ancient Judaism*. Oxford: Littman Library of Jewish Civilization.

Stevens, Wesley. 1995. "Cycles of Time: Calendrical and Astronomical Reckonings in Early Science." In *Cycles of Time and Scientific Learning in Medieval Europe*, 26–49. Aldershot: Variorum.

Strauss, Claudia, and Naomi Quinn. 1997. *A Cognitive Theory of Cultural Meaning*. Cambridge: Cambridge University Press.

Swartz, Marc J. 1991. *The Way the World Is: Cultural Processes and Social Relations among the Mombasa Swahili*. Berkeley: University of California Press.

Symphosius. 1912 (fourth–fifth centuries?). *The Hundred Riddles of Symphosius*, translated by Elizabeth Hickman Du Bois. Woodstock, VT: The Elm Tree Press.

Taub, Liba. 2003. *Ancient Meteorology*. London: Routledge.

Taylor, Charles. 2007. *A Secular Age*. Cambridge, MA: Belknap.

Taylor, Frederick Winslow. 1967 (1947). *Principles of Scientific Management*. New York: Norton.

Thompson, E. P. 1967. "Time, Work-Discipline, and Industrial Capitalism." *Past and Present* 38:56–97.

Toda, M. 1975. "Time and the Structure of Human Cognition." In *The Study of Time II*, edited by J. T. Fraser and N. Lawrence, 314–324. New York: Springer Verlag.

Traweek, Sharon. 1988. *Beamtimes and Lifetimes: The World of High Energy Physicists*. Cambridge: Harvard University Press.

Trinder, John, Stuart Armstrong, Catherine O'Brien, David Luke, and Marion Martin. 1996. "Inhibition of Melatonin Secretion Onset by Low Levels of Illumination." *Journal of Sleep Research* 5:77–82.

Turner, D. C., T. W. Robbins, L. Clark, A. R. Aron, J. Dowson, and B. J. Sahakian. 2003. "Cognitive Enhancing Effects of Modafinil in Healthy Volunteers." *Psychopharmacology* 165:260–269.

Turton, David, and Clive Ruggles. 1978. "Agreeing to Disagree: The Measurement of Duration in a Southwestern Ethiopian Community." *Current Anthropology* 19:585–600.

Tylor, Edward B. 1958 (1871). *Primitive Culture, Volume 1*, New York: Harper and Row.

US Congress, Office of Technology Assessment. 1991. *Biological Rhythms: Implications for the Worker*. Washington, DC: United States Government Printing Office.

Van Dongen, Hans P. A., Greg Maislin, Janet M. Mullington, and David F. Dinges. 2003. "The Cumulative Cost of Additional Wakefulness: Dose-Response Effects on Neurobehavioral Functions and Sleep Physiology from Chronic Sleep Restriction and Total Sleep Deprivation." *Sleep* 26:117–126.

Van Reeth, Olivier, Jeppe Sturis, Maria M. Byrne, John D. Blackman, Mireille L'Hermite-Balériaux, Rachel Leproult, Craig Oliner, Samuel Refetoff, Fred W. Turek, and Eve van Cauter. 1994. "Nocturnal Exercise Phase Delays Circadian Rhythms of Melatonin and Thyrotropin Secretion in Normal Men." *American Journal of Physiology: Endocrinology and Metabolism* 266:E964-E974.

Virgil. 2006 (29 BC). *Georgics*, translated by Peter Fallon. Oxford: Oxford University Press.

von Steuben, Freidrich Wilhelm, Baron. 1779. *Regulations for the Order and Discipline of the Troops of the United States.* Philadelphia: Styner and Cist.

Vygotsky, L. S. 1987 (1934). "Thinking and Speech" In *The Collected Works of L. S. Vygotsky, Volume 1: Problems of General Psychology Including the Volume Thinking and Speech*, edited by Robert W. Rieber and Aaron S. Carton and translated by Norris Minick, 39–285. New York: Plenum Press.

———. 1997 (1925). "Consciousness as a Problem for the Psychology of Behavior." In *The Collected Works of L. S. Vygotsky, Volume 3: Problems of the Theory and History of Psychology*, translated by René van der Veer, and edited by Robert W. Reiber and Jeffrey Wollock, 63–79. New York: Plenum.

———. 1997 (np). "The Instrumental Method in Psychology." In *The Collected Works of L. S. Vygotsky, Volume 3: Problems of the Theory and History of Psychology*, translated by René van der Veer, and edited by Robert W. Reiber and Jeffrey Wollock, 85–59. New York: Plenum.

Wall, John N. 2008. "John Donne and the Practice of Priesthood." In *Renaissance Papers 2007*, edited by Christopher Cobb and M. Thomas Hester, 1–16. Rochester: Camden House.

Wallace, Anthony F. C. 1961. *Culture and Personality.* New York: Random House.

Walpole, Horace. 1798 (1753). "The World by Adam Fitz-Adam, number X, Thursday, March 8, 1753." In *The Works of Horace Walpole, Volume 1*, 159–163. London: G. G. and J. Robinson, and J. Edwards. *Eighteenth Century Collections Online.* Accessed December 6, 2011, from Queens College, Library, CUNY via http://find.galegroup.com/ecco/start.do?prodId=ECCO&userGroupName=cuny_queens.

Walsh, James K., Angela C. Randazzo, Kara L. Stone, and Paula K. Schweitzer. 2004. "Modafinil Improves Alertness, Vigilance, and Executive Function during Simulated Night Shifts." *Sleep* 27:434–439.

Waterhouse, J., D. Minors, S. Folkard, D. Owens, G. Atkinson, I. MacDonald, T. Reilly, N. Sytnik, and P. Tucker. 1998. "Light of Domestic Intensity Produces Phase Shifts of the Circadian Oscillator in Humans." *Neuroscience Letters* 245:97–100.

Wehr, Thomas A. 2001. "Photoperiodism in Humans and Other Primates: Evidence and Implications." *Journal of Biological Rhythms* 16:348–364.

Wehr, Thomas A., Holly A. Geisen, Douglas E. Moul, Erick H. Turner, and Paul J. Schwartz. 1995. "Suppression of Men's Responses to Seasonal Changes in Day Length by Modern Artificial Lighting." *American Journal of Physiology: Regulatory, Integrative and Comparative Physiology* 38:R173–R178.

Weibel, L., and G. Brandenberger. 1998. "Disturbances in Hormonal Profiles of Night Workers during Their Usual Sleep and Work Times." *Journal of Biological Rhythms* 13:202–208.

Weick, Roger S. 1988. *Time Sanctified: The Book of Hours in Medieval Art and Life*. New York: George Braziller.

Wenger, Etienne. 1998. *Communities of Practice: Learning, Meaning and Identity*. Cambridge: Cambridge University Press.

Wertsch, James V. 1998. *Mind as Action*. New York: Oxford University Press.

———. 2007. "Mediation." In *The Cambridge Companion to Vygotsky*, edited by Harry Daniels, Michael Cole, and James Wertsch, 178–192. Cambridge: Cambridge University Press.

Wever, Rütger A. 1979. *The Circadian System of Man: Results of Experiments under Temporal Isolation*. New York: Springer-Verlag.

Whitehead, D. C., H. Thomas, and D. R. Slapper. 1992. "A Rational Approach to Shiftwork in Emergency Medicine." *Annals of Emergency Medicine* 21:1250–1258.

Whorf, Benjamin Lee. 1956 (1940). *Language, Thought and Reality: Selected Writings of Benjamin Lee Whorf*. Cambridge, MA: MIT Press.

Wilks, Ivor. 1992. "On Mentally Mapping Greater Asante: A Study of Time and Motion." *The Journal of African History* 33:175–190.

Winston, Gordon C. 1982. *The Timing of Economic Activities*. Cambridge: Cambridge University Press.

Wolf, Eric. 1982. *Europe and the People without History*. Berkeley: University of California Press.

Wright, Kenneth P, Jr., Claude Gronfier, Jeanne F. Duffy, and Charles A. Czeisler. 2005. "Intrinsic Period and Light Intensity Determine the Phase Relationship between Melatonin and Sleep in Humans." *Journal of Biological Rhythms* 20:168–177.

Yates, Frances. 1966. *The Art of Memory*. Chicago: University of Chicago Press.

Yelvington, Kevin. 1995. *Producing Power: Ethnicity, Gender and Class in a Caribbean Workplace*. Philadelphia: Temple University Press.

York Minster. 1859. *The Fabric Rolls of York Minster*. Durham: Publications of the Surtees Society, Volume 35.

Zeitzer, J. M., D. -J. Dijk, R. E. Kronauer, E. N. Brown, and C. A. Czeisler. 2000. "Sensitivity of the Human Circadian Pacemaker to Nocturnal Light: Melatonin Phase Resetting and Suppression." *Journal of Physiology* 526:695–702.

Zerubavel, Eviatar. 1977. "The French Republican Calendar: A Case Study in the Sociology of Time." *American Sociological Review* 42:868–877.

———. 1979. *Patterns of Hospital Life*. Chicago: University of Chicago Press.

———. 1985. *The Seven Day Circle: The History and Meaning of the Week*. New York: Free Press.

Ziggelaar, August. 1983. "The Papal Bull of 1582 Promulgating a Reform of the Calendar." In *Gregorian Reform of the Calendar*, edited by G. V. Coyne, M. A. Hoskin, and O. Pedersen, 201–239. Vatican: Pontificia Academia Scientiarum.

Index

CPSIA information can be obtained
at www.ICGtesting.com
Printed in the USA
BVHW04s0221230818
525406BV00013B/73/P